T0290945

LACTIC ACID FERMENTATION OF HUMAN EXCRETA FOR AGRICULTURAL APPLICATION

Nadejda Andreev

Thesis committee

Promotor
Prof. Dr P.N.L. Lens
Professor of Environment Biotechnology
UNESCO-IHE Institute for Water Education, Delft

Co-promotors
Prof. B. Boincean
Habil. Dr. of Agricultural Sciences, Research Professor
Research Institute of Field Crops "Selectia"
Balti, Republic of Moldova

Dr M. Ronteltap
Senior Lecturer in Sanitary Engineering
UNESCO-IHE Institute for Water Education, Delft

Other members
Prof. Dr G. Zeeman, Wageningen University & Research
Prof. Dr W.Verstraete, Ghent University, Belgium
Dr Tjaša Griessler Bulc, University of Ljubljana, Slovenia
Prof. Dr J.O. Drangert, Linköping University, Sweden

This research was conducted under the auspices of the SENSE Research School for Socio-Economic and Natural Sciences of the Environment

Lactic acid fermentation of human excreta for agricultural application

Thesis
submitted in fulfilment of the requirements of
the Academic Board of Wageningen University and
the Academic Board of the UNESCO-IHE Institute for Water Education
for the degree of doctor
to be defended in public
on Friday, 29 September 2017at 1.30 p.m.
in Delft, the Netherlands

by Nadejda Andreev
Born in Nisporeni, Boldureşti, Republic of Moldova

CRC Press/Balkema is an imprint of the Taylor & Francis Group, an informa business

Although all care is taken to ensure integrity and the quality of this publication and the information herein, no responsibility is assumed by the publishers, the author nor UNESCO-IHE for any damage to the property or persons as a result of operation or use of this publication and/or the information contained herein.

Published by:

CRC Press/Balkema

Schipholweg 107C, 2316 XC, Leiden, the Netherlands

Pub.NL@taylorandfrancis.com

www.crcpress.com – www.taylorandfrancis.com

ISBN 978-1-138-04989-5 (Taylor & Francis Group)

ISBN 978-94-6343-057-9 (Wageningen University)

DOI: https://doi.org/10.18174/401835

Dedication

This thesis is dedicated to the research community interested in resource oriented sanitation.

Aknowledgements

I would like to express sincere gratitude to my promotor Prof. Dr. ir. Piet Lens and to my co-promotors Dr. Mariska Ronteltap and Prof. Boris Boincean for scientific support, guidance, encouragement and critical interpretation of the data, who enabled me to earn an open minded research thinking and helped in a systematic way of finding answers to questions.

I am also very grateful to prof. Elena Zubcov, Nina Boichenco and Natalia Borodin from the Laboratory of Hydrobiology and Ecotoxicology, Institute of Zoology of Moldova; Dr. Lidia Bulat from the Department of Sustainable Farming Systems of Crop Research Insitute, Selectia, Baltsy, Moldova; Dr. Larisa Cremeneac from the National Insitute of Zootechny and Veterinary Medicine, Moldova as well as Svetlana Prudnichonok and Olga Coteţ from the Laboratory of Sanitary Microbiology, National Centre of Public Health, Moldova for the research advise and support during crucial parts of my field and laboratory research experiments. Many thanks to Christopher Canaday, biologist at the Fundación Omaere (Puyo, Ecuador) for proofreading the thesis and providing valuable comments for its improvement.

Special thanks to my relatives Elena, Vladimir, Nadine, Vasile and Ana Vrabie, Semion Varaniţă as well as Vera and Boris Suciu who offerred valuable help during the field work. I am also indebted to Sergiu Andreev who provided support in graphical design of the thesis. I am also very thankful to my beloved children, Alina and Sandu for their patience and understanding during my absence for field work or when being abroad. I give all the respect to my departed mother and father, who gave me a lot of encouragement during the start of my research.

Nadejda Andreev
Delft, 24 March, 2017

Table of Contents

Thesis summary

Management of human excreta is one of the most pressing global issues. The application of cost-effective, environmentally friendly treatment methods and agricultural reuse of excreta are particularly important in the context of global nutrient and resource challenges, a rapid depletion of soil nutrients and organic matter, eutrophication and scarcity of freshwater resources, exacerbated by the population growth and climate change. This study focuses on the use of lactic acid fermentation for the treatment of human faeces and urine for agricultural reuse.

The efficiency of lactic acid fermentation and thermophilic composting versus lactic acid fermentation and vermi-composting on pathogen removal and the post-effects on seed germination (radish) and plant growth (tomatoes) were assessed in pot experiments. The main effects of lacto-fermented faeces and bio-waste supplemented by biochar were also studied during a two-year experiment under the conditions of a clay-loamy cernoziom in the Central part of Moldova using corn, *Zea mays* L, as test crop. Urine lactic acid fermentation was also investigated under laboratory conditions and real field conditions in a functional household urine diverting dry toilet (UDDT) storage tank in the vicinity of Chisinau (Moldova).

The research results revealed that lactic acid fermentation combined with thermophilic composting reduced concentration of the coliforms, *Escherichia coli*, *Enterococcus faecalis* and *Clostridium perfringens* to below the detection limit, while lactic acid fermentation combined with vermi-composting reduced the coliform concentration to 5 log CFU g^{-1} only. Lactic acid fermentation of the mix of faeces and bio-waste with the addition of lactic acid bacteria and molasses followed by thermophilic composting led to a rapid increase in sanitization temperature above 55° C, without the need of turning and aeration. The overall composting period of lactic acid fermentation combined with thermophilic composting was 33 days compared to 130 days for combined lactic acid fermentation with vermi-composting. The germination index of radish seeds fertilized by compost after lactic acid fermentation combined with thermophilic composting was higher than that obtained after lactic acid fermentation combined with vermi-composting (90 versus 84 %, respectively). Moreover, significantly bigger average fruit weight and total amount of biomass per tomato plant ($p<0.05$) were obtained after compost amendment compared to vermi-compost and the unamended control.

The difference of means by the Dunnet test showed that the mix of lacto-fermented faeces and bio-waste, supplemented by urine charged biochar has increased significantly ($p<0.05$) the corn growth, yield and yield components compared to the stored urine, faeces and cattle manure. The yield of corn after fertilization of mineral fertilizer during the first year was significantly lower than the investigated fertilizing mix, but during the second year, it was not significantly different due to different weather (precipitation) conditions. The lacto-fermented mix of faeces and bio-waste, supplemented by biochar also significantly improved the soil bulk density, soil mobile potassium and soil nitrate content (the latter only in comparison to the control and stored urine). The beneficial effects of lacto-fermented faeces and bio-waste, supplemented by urine charged biochar on yield components might be attributed to potential prevention of nitrate leaching by the biochar component in the root zone.

The application of a lactobacilli solution of sauerkraut containing molasses and water to fresh urine contributed to an efficient reduction of the pH to < 4-4.5 and of the ammonium content by 1/3, while maintaining a high concentration of viable LAB (7.3 CFU ml^{-1}) compared to untreated, stored urine. Moreover, the odour strength was reduced twicely after urine lactic acid fermentation. Also the seed germination was enhanced after application of lacto-fermented urine, showing a potentially higher fertilizing value than the stored urine.

This study showed that lacto-fermented faeces and bio-waste, supplemented by urine charged biochar can serve as a potential suitable soil conditioner. Besides, it showed that lacto-fermented urine can be a potential bio-fertilizer for improving soil quality and crop yield. The results of this thesis can be useful for improving the resource oriented potential of urine diverting dry toilets.

Samenvatting

De adequate behandeling van menselijke uitwerpselen is een van de meest uitdagende mondiale kwesties. De toepassing van kosteneffectieve, milieuvriendelijke behandelingsmethoden en agrarische hergebruik van uitscheidingsproducten zijn bijzonder belangrijk in het kader van de globale uitdagingen op het gebied van nutriënten; uitputting van de bodem met betrekking tot nutriënten en organische stof; vervuiling, eutrofiëring en schaarste van zoetwaterreserves, welke versterkt worden door de bevolkingsgroei en klimaatverandering. De huidige studie richt zich op de inzet van lacto-fermentatie voor de behandeling van menselijke uitwerpselen en urine voor hergebruik in de landbouw.

De efficiëntie van lacto-fermentatie en thermofiele compostering versus lacto-fermentatie en vermi-compostering op pathogeenverwijdering en vervolgens de effecten op zaadontwikkeling (radijs) en plantgroei (tomaten) zijn beoordeeld aan de hand van een 24 maands onderzoek, zowel in de volle grond als in plantenbakken.

De belangrijkste effecten van lacto-gefermenteerde feces en groen afval aangevuld met biochar werden ook bestudeerd tijdens een experiment van twee jaar in een leemklei-cernoziom bodem in het centrale deel van Moldavië met behulp van maïs (*Zea maïs* L.) als testgewas. Urine lacto-fermentatie werd eveneens onderzocht onder laboratoriumomstandigheden en echte veldomstandigheden met urine opgevangen in een droog scheidingstoilet (urine diverting dehydrating toilet, UDDT) in de nabijheid van Chisinau, Moldavië.

Resultaten van het onderzoek toonden aan dat middels lacto-fermentatie in combinatie met thermofiele compostering het aantal coliforme bacteriën (*Enterococcus faecalis*, *Escherichia coli* en *Clostridium perfringens*) kon worden teruggebracht tot onder de detectielimiet; middels lacto-fermentatie in combinatie met vermi-compostering slechts tot 5 log CFUg^{-1}. Lacto-fermentatie van de mix van feces en bio-afval met de toevoeging van melkzuurbacteriën en melasse gevolgd door thermofiele compostering leidt tot een verhoging van de temperatuur tot boven 55°C, zonder mengen of beluchten. De algehele composteringsperiode van lacto-fermentatie gecombineerd met thermofiele compostering was 33 dagen; voor gecombineerde lacto-vergisting met vermi-compostering bedroeg dit 130 dagen. De kiemindex van radijszaden bemest met compost na lacto-vergisting plus thermofiele compostering was hoger dan die na lacto-vergisting plus vermi-compostering (90 tegenover 84%, respectievelijk). Bovendien werd er per tomatenplant een significant grotere gemiddeld gewicht per vrucht en per plant verkregen ($p < 0.05$) dan na het toedienen van vermicompost of geen compost (controle).

Het verschil van gemiddelden werd bepaald middels de Dunnet test. Hieruit kwam naar voren dat groei, rendement en opbrengst van de maïs significant toenam ($p < 0.05$) na toevoeging van lacto-gefermenteerde feces en bio-afval aangevuld met urine en biochar, ten opzichte van gehydrolyseerde urine, feces en mest. De opbrengst van maïs na bemesting met minerale meststof was aanzienlijk lager tijdens het eerste productiejaar maar niet significant anders dan de onderzochte bemestingsmix tijdens het tweede productiejaar. Lacto-gefermenteerde mix van feces en bio-afval, aangevuld met biochar, heeft de bulkdichtheid, mobiele kalium en het nitraatgehalte van de bodem verbeterd (het nitraatgehalte in vergelijking met de controle en de opgeslagen urine). De gunstige effecten van lacto-gefermenteerde feces en bio-afval aangevuld met urine en biochar op de plantendelen kunnen worden toegeschreven aan potentiële preventie van nitraatuitspoeling in de wortelzone door biochar component tijdens droge perioden, gevolgd door regen.

De toepassing van een oplossing van lactobacilli afkomstig van zuurkool, melasse en water met verse urine droeg bij aan een verlaging van de pH naar minder dan 4-4,5 en van het ammoniumgehalte met 30%, met behoud van een hoge concentratie van levensvatbare LAB (7,3 ml CFU^{-1}) in vergelijking met onbehandelde, gehydrolyseerde urine. Daarnaast droeg lacto-gefermenteerde urine bij aan een verminderde waargenomen geur en had het een positieve invloed op de kiemkracht van zaad.

Deze studie toonde aan dat lacto-gefermenteerde feces en bio-afval, aangevuld met urine doordrenkte biochar, kunnen dienen als een potentiële bodemverbeteraar, en lacto-gefermenteerde urine als potentiële bio-meststof voor de verbetering van de kwaliteit van de bodem en de opbrengst van gewassen. De resultaten van deze thesis kunnen nuttig zijn voor verbetering van het hergebruikspotentieel van de grondstoffen uit droge scheidingstoiletten (UDDTs).

Chapter 1. General Introduction

1.1 Research justification

Modern society is faced with several challenges regarding the disposal and treatment of human excreta. Water and resource limitations in conventional sanitation systems are becoming nowadays of a concern due to urbanization and population growth, negative effects on freshwater and land resources as well as aggravating climate change. In spite of technological development in the field of sanitation, still a high proportion of the generated wastewater is discharged into the environment receiving a poor or even no treatment at all (Sato *et al.*, 2013). Many developed and developing countries bear significant costs for maintaining and upgrading the sanitation infrastructure and treatment facilities (Schertenleib, 2005a; Zimmer and Hofwegen, 2006; Kone, 2010).

The scarcity of water and limitations of financial resources or political willpower force a majority of the population of Eastern and Central Europe, Africa, South, South-East and East Asia as well as Central America who live in rural settlements to use unsewered, environmentally polluting latrine types of sanitation (Bodik, 2007; Kone, 2010). If not properly managed, these sanitation systems can lead to high risks for environmental and public health in the crowded areas, those with a high water table, those that are periodically flooded, hard to excavate or in sensitive coastal zones (Winblad and Simpson-Hebert, 2004).

Also modern agriculture is still based on intensive use of nonrenewable sources of energy (oil, natural gases and coal) and mineral fertilizers, especially nitrogen. Recycling of organic matter and nutrients to the soil is limited. While high amounts of nutrients, with a fertilizing potential, are discharged via excreta or wastewater, the global phosphorous resources are depleting (Cordell and White, 2011) and the soils are threatened due to a decrease of their natural fertility through so-called "nutrient and organic matter mining". For example, in many developing countries and those with economies in transition, where farmer survival depends heavily on agriculture and land, the fertilizing input is very low due to high prices or the absence of mineral fertilizers. Thus, soils are drained of organic matter and nutrients making them more prone to degradation under natural conditions or by man made factors. A similar phenomenon happens to many European soils, where approximately half of the soils have reached a low or very low organic matter content. In Southern Europe for instance, a high proportion of intensively used agricultural soils reach a critical threshold value for soil organic carbon (2%) (Rusco et al., 2001).

A "material-flow-oriented recycling sanitation", also called "resource oriented" or "productive sanitation" approach is seen as an effective solution for recycling organic matter and nutrients to the soil (Murray and Buckley, 2010). This new paradigm shifts the idea from disposal of excreta to containment and recycling. This fits into the idea of a circular economy, where environmental and economic gains are obtained from the recycling of nutrients and organic matter. Recycling of nutrients from excreta to agriculture can reduce the costs from the use and production of chemical fertilizers. Of a particular importance is the case of phosphorus fertilizers, which prices are excalating at the global level due to shrinkage of rock resources and raise in the costs for their extraction and transportation. For example, in 2008 a steady increment by 800 percent was encountered for phophate rock due to boosting of oil prices, cultivation of biofuels, more demand for fertilizers caused by higher meat consumption and short term scarcity of phosphorus rock resources on the global market. That led to rise in food prices that hit poor countries very hard (Cho, 2013).

Reuse of excreta is one of the potential ways to supplement the scarce phosphorus resources. Estimations from USA show that recovery and recycling of phosphorus from excreta, food waste and animal manure can return up to 37 % of annual crop requirements (Carlson, 2016). The safe recycling of these nutrients, however, does require effective sanitisation technologies. Working with source separated urine and faecal matter asks for alternative methods to more conventional biological wastewater treatment technologies, for which we need to reorient ourselves by either looking for innovative approaches or by applying known technologies in a novel fashion.

One promising example from the past is the application of charred organic matter in combination with other organic material, which was proven to be a viable strategy for replenishing soil nutrients and organic matter deficiencies. In the past, highly fertile, artificially created soils, "terra preta" or "black anthrosols" have been distributed over South America, Australia, West Africa and Northern Europe (Downie *et al.*, 2011; Frausin *et al.*,

2014; Wiedner *et al.*, 2015a). These soils arose as a result of intentional or unintentional deposition of organic (e.g. left-over cooking charcoal, human and animal excreta and plant material) and inorganic waste (e.g. ceramic shreds) on the top of infertile soil (Glaser and Birk, 2012a; Giani *et al.*, 2014). After the discovery of terra preta soils and its potential was realized, extended research was initiated on investigating the role of biochar in increasing soil fertility as well as the re-application of ancient practices of waste management and soil enrichment in the field of sanitation and agriculture (Kawa and Oyuela-Caycedo, 2008; Lehmann and Joseph, 2009b; Novotny *et al.*, 2009; Otterpohl and Buzie, 2011; Smetanova *et al.*, 2012; Otterpohl and Buzie, 2013; Windberg *et al.*, 2013; Schuetze and Santiago-Fandino, 2014; Yemaneh, 2015).

Lacto-fermented and composted human excreta can provide a range of useful products of natural origin for agriculture, e.g. bio-fertilizers such as lacto-fermented urine or soil conditioners (compost obtained after lactic acid fermentation combined with thermophilic or vermi-composting). The obtained end products are bio-degradable and have a lower detrimental impact on humans and plants than chemical fertilizers. The microbial inocula applied can be obtained from locally available resources such as sauerkraut, rice wash or milk products (Aliyu & Muntari, 2011; Factura *et al.*, 2010). Lactic acid fermentation is a low cost technique that employs limited handling and makes use of simple equipment and limited resources (Erickson *et al.*, 2004).

Terra preta sanitation is an approach where human excreta is treated and recycled to agriculture and horticulture in an original way via combined lactic acid fermentation as a primary treatment step, followed by vermi-composting as a secondary post-treatment step. Recently, thermophilic composting was proposed as a post-secondary step to efficiently recycle the nutrients and organic matter as well as to reduce the pathogen contamination (Andreev *et al.*, 2017). Biochar supplementation at the end of the lactic acid fermentation or at the thermophilic or vermi-composting stage increases the stability of carbon matter and reduces the nitrogen loss via volatilization. Lactic acid fermentation of urine and faeces diminishes the odour emissions and contributes to their stabilization. Faeces lactic acid fermentation after the collection phase was combined with other types of bio-waste, such as high moisture and carbon rich waste from the food industry, e.g. from diary or beverage industry or the kitchen scraps. These wastes are easily degradable and can produce high emissions and pungent odours (Kosseva, 2013). By combining lactic acid fermentation with thermophilic composting, the health risks during faeces handling can be reduced as temperature increases occur in the substrate with passive aeration only. Apart from faeces, also lactic acid fermentation of human urine can be considered as it reduces the pH and decreases the ammonia volatilization, thus potentially enhancing the fertilizing value of urine (effects on germination) and reducing odour emissions. Moreover, lacto-fermented urine may have a better effect on germination owing to the metabolites lactic acid bacteria (LAB) produce (e.g. organic acids and hydrogen peroxide) and suppressing of a wide range of phytopathogens (Limanska *et al.*, 2015), while having a stimulatory effect on plant growth by triggering plant systemic resistance and enhance production of plant growth regulators (Abel-Aziz *et al.*, 2014).

1.2 Scope of research

This PhD thesis assessed the suitability of lactic acid fermentation for increasing the fertilizing value of human excreta (Figure 1.1). The efficiency of lactic acid fermentation and thermophilic composting versus lactic acid fermentation and vermi-composting on pathogen removal and the post-effects on seed germination (radish) and plant growth (tomatoes) were assessed in pot and field experiments. The main effects of lacto-fermented faeces and biowaste supplemented by biochar on soil quality of a clay-loamy cernoziom in the central part of Moldova as well as on the growth and yield of corn (*Zea mays* L.) were studied during a two-year field experiment. The field research was carried out in randomized experimental plots of 60 m², with three replicates on a total area of 0.15 ha. Lactic acid fermentation of faeces and the bio-waste mix was performed in fermentation pits and closed plastic barrels. Thermophilic composting was carried out in a metallic insulated box and vermi-composting in a windrow of 1 m², for a period of four months. Earthworms (*Eisenia foetida*) were acquired from the Institute of Biotechnologies in Zootechny and Veterinary Medicine (Maximovca vg., Anenii Noi district, Moldova).

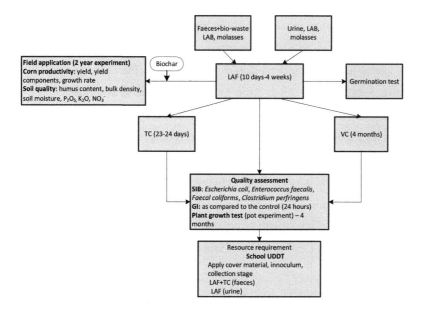

Figure 1.1 Main experimental stages for faeces and urine treatment via lacto-fermentation and post-treatment. LAF - lacto-fermentation, LAB - lactic acid bacteria, TC - thermophilic composting, VC - vermi-composting, SIB - sanitization indicator bacteria, GI - germination index.

1.3 Thesis outline

This thesis is organized into 6 chapters: the introductory part, literature review and 3 research chapters as well as one discussion chapter. The introductory part gives background information of the study topic, scope and outline of the research. The second chapter is a literature review and describes the main nutrient and resource challenges faced by conventional sanitation, the rate of soil degradation and the loss of organic matter in the soil as well as the need for sustainable build-up of soil organic matter, including the application of combined excreta and biochar, the fertilizing value of human excreta as well as advantages and disadvantages of the use of biochar in agriculture. Furthermore, the chapter discusses the challenges related to agricultural reuse of excreta such as hygienization aspects, loss of nutrients, carbon matter and odour issues as well as the potential for lactic acid fermentation to overcome these challenges. The chapter also addresses the limitations of lactic acid fermentation, post-treatment of faeces via thermophilic composting and vermi-composting, the role of biochar supplementation as well as agricultural effects of lacto-fermented excreta.

The third chapter evaluates the efficiency of lactic acid fermentation and thermophilic composting versus lactic acid fermentation and vermi-composting for pathogen removal and the effects on plant germination and growth. The results obtained from this chapter showed that combined lactic acid fermentation with thermophilic composting are more efficient in pathogen removal and plant growth than combined lactic acid fermentation with vermi-composting. In the fourth chapter, the effects of lacto-fermented faeces and biowaste, supplemented by biochar on soil quality (soil nitrate, phosphorus potassium and humus content, moisture content as well as bulk density) and corn productivity (growth, yield and yield components) was examined in a two year field experiment. A comparison was made with no fertilization, lacto-fermented mix without biochar, stored human faeces, cattle manure, urine as well as nitrogen, phosphorus and potassium mineral fertilizers. The lacto-fermented mix of faeces and bio-waste, supplemented by biochar significantly improved plant height ($p < 0.05$) compared to all fertilizers during the first production year and compared to the control, stored faeces and vermi-compost during the second year. This fertilizer also achieved a significantly higher corn yield compared to all other fertilizers during the first and second production year, except for the lacto-fermented mix without biochar and the mineral fertilizer, which showed no significant yield difference ($p>0.05$). It also reduced the bulk density of the soil during both years and increased the soil potassium content during the first production year. The yield components (rows per ear and kernels per ear) were improved significantly when supplied with the lacto-fermented mix of faeces and bio-waste, supplemented by biochar that might have attributed to potential prevention of nitrate leaching in the root zone during dry spells, followed by rain flushes.

Chapter 5 discusses the potential of the improvement of the fertilizing value of urine and the prevention of a pH increase and ammonia loss by urine lactic acid fermentation using low cost LAB inoculum. The results showed that the urine lactic acid fermentation has led to effective acidification to pH < 4 and a reduction by 1/3 of the ammonium content compared to the stored urine. Moreover, lacto-fermented urine reduced twice the perceived odour strength and improved seed germination, thus showing a potentially higher fertilizing effect than untreated, stored urine.

Chapter 6 examines the potential applications of the research results in urine diverting dry toilets (UDDT) in Moldova for the primary and secondary treatment of both the urine and faeces fractions. The aspects on product quality requirements for potential full applications and the input materials are given. Furthermore, an entrepreneurial model is proposed for the application of compost obtained from lactic acid fermentation of faeces, kitchen and garden waste. In this model, nutrient rich, lacto-fermented urine is used for the irrigation of a short rotation copice (willow), grown for heating of schools. Faeces treated via combined lactic acid fermentation and thermophilic composting is applied as growth medium for ornamental plants. Potential revenue streams from sale of products (willow plantation and ornamental plants) are highlighted.

Chapter 2. Increasing the agricultural value of human excreta by lactic acid fermentation, composting and biochar addition (literature review)

This chapter is based on:

Nadejda Andreev, Mariska Ronteltap, Boris Boincean, Piet N.L. Lens. Lactic acid fermentation of human excreta for agricultural application (under review).

Abstract

Studies show that source separated excreta have a good fertilizing potential for improving crop production and soil quality, the encountered effects are similar or even exceed those of mineral fertilizers. The main challenges with agricultural application of excreta are the high pathogenic risks (especially of the faeces fraction), increased loss of nutrients and odour emissions. The use of lactic acid fermentation of excreta for improving its resource-oriented value as well the post-treatment stages via vermi-composting and micro-aerobic composting is reviewed. The review shows that lactic acid fermentation of excreta can increase its agricultural value by reducing the amount of pathogens, minimizing the nutrient loss and inhibiting the production of malodourous compounds. While pathogens as *Enterobateriacea, Staphylococcus* and *Clostridium* can be reduced by 7 log CFU g^{-1} during 7-10 days of fermentation, *Ascaris* may not be always efficiently removed. Direct application of lacto-fermented faeces to agriculture may be constrained by incomplete decomposition, high content of organic acids or insufficient hygienization. Post treatment by biochar addition, vermi-composting or thermophilic composting will contribute to stabilization and sanitizing of the material. The pot and field experiments on soil conditioners obtained by lactic acid fermentation and post treatment steps by composting and biochar supplementation demonstrate an increased crop yield and growth as well as improved soil quality in comparison to the unfertilized control or other types of fertilizers (e.g. mineral fertilizer, stored faeces, urine or cattle manure).

2.1 Introduction

Global population growth, intense urbanization and economic development as well as climate change increase competition over water, energy and land resources. While there is recognition of the need to raise the food production for the growing population, there is also a concern of the limitations of land and freshwater resources, which are undergoing worldwide degradation (Hoff, 2011; Turral *et al.*, 2011). Energy can be substituted from renewable resources, while water and soil resources have no substitute. Therefore, it is important to prevent their further degradation and where possible restore their quality.

Soil is lost at a 10-40 times faster rate than it is regenerated, approximately one quarter of all cultivated land is undergoing degradation (Pimentel, 2006; Bai *et al.*, 2008). For this reason, farmers must spend more on energy, without a corresponding increase in crop yield, thus negatively affecting the livelihood of poor populations, especially in the developing countries of Africa, Asia and Latin America, whose economies depend on agriculture. Erosion and desertification are increasing throughout the world, which declines agricultural production in India, Pakistan, Nepal, Iran, Jordan, Lebanon and Israel (Pfeiffer, 2006).

Soil depletion and imbalances are a widespread problem in Africa and Asia (Tan *et al.*, 2005). The loss of nutrients and organic matter is also a concern in Europe, where approximately 45% of soils have reached a low or very low organic matter content of 0-2% (Rusco *et al.*, 2001). This depletion is particularly intense in the Mediterranean region, Southern and Eastern Europe, as well as in some countries in Western Europe (Jones *et al.*, 2012; Krupenikov *et al.*, 2011; Virto *et al.*, 2014).

The loss of soil organic matter leads to a reduction in soil fertility as well as its structure, water holding capacity and biological activity of the soil. Soil organic matter is lost when organic matter inputs are not replenished to the soil during cultivation (Vlek, 1997). Along the history of crop fertilization, farmers gradually replaced organic fertilizers with mineral fertilizers as the latter were more convenient to store, transport or spray (Schröder, 2014). In many parts of the world, increased application of nitrogen fertilizers was advocated and subsidised (Mulvaney *et al.*, 2009). Long term application of solely mineral fertilizers can reduce soil organic carbon content (Boincean *et al.*, 2014) or cause other adverse effects, such as acidification (Guo *et al.*, 2010).

Freshwater resources around the globe are also faced by degradation and scarcity issues. For a number of countries in Africa and the Middle East, water is becoming physically scarce, while for others (e.g. Latin America) it is economically scarce (Rijsberman, 2006). Degradation of water bodies and water-related ecosystem services is a widespread global problem (FAO, 2011). Sanitation is a major consumer and pollutant of water resources (Gleick, 2003). Large volumes of drinking water are used to transport excreta, generating enormous amounts of wastewater which cannot be fully cleaned with existing conventional technology. This results in significant amounts of nutrients and organic matter being discharged into surface waters, especially since 90% of the wastewater is released without treatment. Reusing excreta in agriculture returns nutrients to the soil and reduces the pollution of freshwater resources and, thus, the energy costs required for its treatment. Reusing excreta to agriculture will allow replacing or complementing the mineral fertilizers and reduce the pollution of freshwater resources and thus the energy costs required for its treatment.

Phosphorus reuse from sanitation is of particular interest, considering the projected increase in the price and availability of phosphorous fertilizers due to depletion of mineral reserves, the uneven geopolitical distribution and the increasing energy costs for their mining, processing and extraction (Cordell and White, 2011). If only the phosphorous content from urine and faeces could be recovered and reused, it would supply 22 % of the current global phosphorous demand in agriculture (Mihelcic *et al.*, 2011).

Prehistoric human societies have created long-lasting rich fertile soils by integrating their excrement, biochar and other substances into the ground (Lehmann *et al.*, 2003). This occurred in the Amazon terra preta or Amazonian dark earths (Glaser and Birk, 2012; Woods, 2003), Northern Europe (Wiedner *et al.*, 2015 a), Australia (Downie *et al.*, 2011), and West Africa (Frausin *et al.*, 2014). This ancient practice of forming so-called terra preta soils highlights the potential for application of excreta and biochar for improving the soil fertility.

This ancient practice of terra preta soils highlights the potential for modern application of excreta and biochar for improving the soil fertility. A resource-oriented approach, named "terra preta sanitation system" has recently been developed (Otterpohl and Buzie, 2011; Otterpohl and Buzie, 2013; Schuetze and Santiago-Fandiño, 2014; Windberg *et al.*, 2013), which treats excreta by two combined processes: lactic acid fermentation (LAF), followed by composting (usually worm composting). Biochar is also applied to reduce the nutrient losses and obtain stable organic soil conditioners (Bettendorf *et al.*, 2014; Glaser, 2015; Yemaneh *et al.*, 2014). Lactic acid fermentation contributed to controlling foul odour (Yemaneh *et al.*, 2014) and suppresses the growth of pathogenic bacteria (Scheinemann *et al.*, 2015). It also shortens the required stabilization time during the subsequent vermi-composting stage (with earthworms), where further pathogen reduction, fragmentation and aeration occur in the faeces (Otterpohl and Buzie, 2011; Otterpohl and Buzie, 2013)

Different aspects of terra preta sanitation have been investigated so far, e.g. integration of terra preta sanitation in different sanitation systems (Schuetze and Santiago-Fandino, 2014; Bettendorf *et al.*, 2015), identification of effective LAB for excreta treatment; optimization of carbohydrate sources for efficient hygienization and odour mitigation (Yemaneh *et al.*, 2012; Yemaneh *et al.*, 2014; Yemaneh, 2015). Research on the effects of excreta treated via terra preta sanitation on plant soil systems is rather scarce.

This chapter evaluates the efficiency of LAF, followed by vermi-composting and microaerobic composting, in the treatment of excreta to improve its agricultural value, reduce pathogens, control odour, and generate a long-lasting, stable product. The role of biochar for avoiding nutrient loss and contributing to the formation of humus was highlighted and the potential fertilizing effects of excreta treated via LAF combined with composting and biochar addition were overviewed. Due to the scarcity of peer-reviewed publications on LAF of excreta, articles on human food and animal feed were also reviewed. For a better appreciation of the potential applications of this approach, a urine-diverting dry toilet was taken as an example, in which the main fermentation conditions and additives required for their optimization at different steps were analyzed separately for the urine and the faeces.

2.2 Nutrient and resource challenges in sanitation

Conventional water-based sanitation aims to increase hygiene, but it is not applicable in all the societies of the world. Clean drinking water is used to transport excrement and is then treated via an energy-intensive process, but is never 100% clean again (Figure 1, I). This is inappropriate, especially in regions with scarcity of freshwater or energy resources, which are growing due to droughts caused by global climate change (Hanjra and Qureshi, 2010). An issue of concern is the pollution of freshwater resources that increases the costs of water treatment.

Freshwater pollution increases the costs of potable water treatement, but the wastewater that contaminates it contains numerous valuable resources, such as nutrients and organic matter from the food people eat (Figure 2.1, II), in addition to constituents of concern, e.g. pathogens and trace amounts of pharmaceuticals, hormones and pesticides (Figure 2.1, III). Wastewater is collected and transported to the treatment facilities (Figure 2.1, IV), where the pollutants are only partly removed (Muga and Mihelcic, 2008), in addition to getting mixed with industrial wastewater and stormwater (Figure 2.1, V) (Tchobanoglous *et* al., 2003). Even with advanced treatment such as microfiltration and reverse osmosis (Watkinson *et al.*, 2007), micro-pollutants (e.g. pesticides, phenol compounds, heavy metals, pharmaceuticals and personal care products) cannot be fully removed during the treatment processes (Tchobanoglous *et al.*, 2003). They thus present a high risk to aquatic organisms (Cirja *et al.*, 2008) and people living downstream. At the same time, valuable resources like nitrogen and phosphorus are not efficiently recycled (Katukiza *et al.*, 2012; Rosemarin, 2010;). While part of them are recovered during the treatment process and reused in agriculture (Figure 2.1, VI), another part is released into aquatic ecosystems, causing, along with agricultural runoffs (Figure 2.1, VIII), eutrophication and impairment of water quality (Figure 2.1, VII) (Tan *et al.*, 2005; Jianyao *et al.*, 2010; Syers *et al.*, 2011).

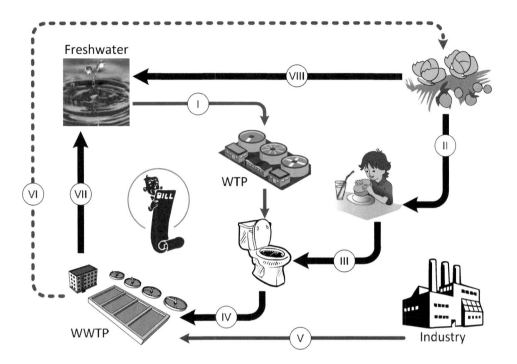

Figure 2.1 Resource and nutrient challenges in conventional sanitation. *A major resource challenge is that treated water (I) is used as an excreta dilution and transportation medium. Valuable resources (e.g. nutrients and organic matter) (II) enter our body through food and are excreted (III), being flushed away into the wastewater collection systems (IV). There, they are mixed with industrial wastewater (V) and treated at wastewater treatment plants. Only a minor part of the nutrients and carbon matter is recycled to food production systems (VI), while the major part is discharged into aquatic ecosystems (VII), causing eutrophication and increasing the costs of water treatment. WTP - water treatment plant; WWTP - wastewater treatment plant.*

One limitation of conventional waterborne sanitation is the high maintenance and operation cost (Schertenleib, 2005b). Most developed countries are already confronted with significant costs for modernizing and upgrading their sanitation infrastructure, but in developing countries, the scarcity of financial and human resources often renders large wastewater treatment facilities dysfunctional only after a few years of operation. For example, the government of Thailand has invested during the 1990s 1.5 billion USD in wastewater and treatment facilities, but only 40 % were functioning properly shortly after their installation (Schertenleib, 2005). Concern also rises over the limitations of the land suitable for landfilling of sewage sludge resulting from wastewater treatment plants (Jiménez *et al.*, 2010).

Most of the urban and peri-urban population of Africa, Asia and Central America as well as rural populations from Eastern Europe are relying on pit latrines as primary means of sanitation. This system is affordable for many people in water scarce areas and those with limited financial resources (Kvarnström *et al.*, 2006; Morella *et al.*, 2009). They, however, contribute to heavy groundwater pollution and contamination of the soil and water with nutrients and pathogens, thus increasing the health risk of those using groundwater as a drinking water source (Kimani-Murage and Ngindu, 2007; Graham and Polizzotto, 2013). In addition, pit latrines are not suitable for crowded areas, sensitive coastal areas, soils with rocky grounds or those with a high water table or periodically flooded (Winblad, 1997; Hurtado, 2005). The potential increase in intensity of floods and rise of the groundwater level as a result of climate change can make this type of sanitation less suitable for some areas, especially in the coastal zones (Graham and Polizzotto, 2013).

2.3 Applying excreta and biochar to agriculture

2.3.1 Fertilizer value of human urine and faeces

If appropriately treated, human urine and faeces have a good potential for substitution or complementing mineral fertilizers and soil conditioners. Resource oriented sanitation systems are efficient for recycling nutrients and carbon matter of human excreta (Larsen *et al.*, 2009; Murray and Buckley, 2010). In some source separating sanitation systems, urine, which the main source of plant available nutrients, is diverted by a separate outlet to a storage tank and the faeces fraction is collected in special collection chambers or containers, to which a bulking agent (e.g. sawdust or ash) is generally added for dehydration, smell reduction and decomposition (Anand and Apul, 2014). After storage or other additional treatment for hygienization and stabilization, urine and faeces can be applied to crop production systems, thus ensuring a nutrient closed loop approach.

In human urine, nitrogen, phosphorous, potassium and sulphur are present in plant accessible, ionic forms (Jönsson *et al.*, 2004). Additional components with fertilizing value contained in urine are calcium, magnesium and micro-elements, e.g. B, Cu, Zn, Mo, Fe, Co and Mn (Kirchmann and Pettersson, 1994; Rodushkin and Ödman, 2001). The heavy metal content is usually low in human urine (Kirchmann and Pettersson, 1994). It was calculated that the application of urine can be economically more feasible than the production and use of mineral fertilizers even when this is transported 100-200 km away (Johansson *et al.*, 2000). Pot experiments on barley with ^{15}N and ^{32}P labelling phosphorus and nitrogen crop uptake from human urine which was comparable to the soluble nitrogen and phosphorus uptake from mineral fertilizers (Kirchmann and Pettersson, 1994).

Faeces account for a smaller quantity than the urine (approx.1/10) and nutrients, especially nitrogen, are organically bound (Jönsson, 2003; Winker *et al.*, 2009). As the amounts of nutrients, such as N, K and S (Rose *et al.*, 2015) would probably cover only a part of the crops' needs, the faecal derived compost could be used primarily as a P-rich organic soil supplement and as a soil conditioner (Winker *et al.*, 2009). For improving the fraction of organic carbon and aeration (Strauss *et al.*, 2003), mixing of faeces with organic waste and carbon rich bulking material is recommended, which also enhances the thermophilic composting process important for hygienization and reuse (Gotaas, 1956; Niwagaba, 2009; Winker *et al.*, 2009).

There have been many studies on the effect of source separated urine on different crops and soils, but limited research exists on the effect of faeces or combined faeces and urine on crops and soils (Table 2.1). Urine application had a beneficial influence on different crops such as okra, cabbage, tomatoes and cucumber by increasing crop nutrient uptake, yield or pest resistance, these effects being similar or even higher than those of mineral fertilizers

Lactic acid fermentation of human excreta for agricultural application

(Heinonen-Tanski *et al.*, 2007; Pradhan *et al.*, 2007; Pradhan *et al.*, 2009; Akpan-Idiok *et al.*, 2012.

Table 2.1 Overview of the effects of source separated urine and faeces on crops and soil quality

Experimental conditions	Fertilizer	Crop type	Effects	Length of experiment	Ref.
Sandy loamy soil. Fertilizer rate: urine - 0,10, 15 and 20 m3L-1, 400 kg N ha-1, 2 weeks after transplanting Greeh. trial: plastic pots field experim. - plots 20x20m 50 plants/plot	NPK and control[2,] UDT[3] urine,	Okra[1]	Increase in the nutrient uptake, growth and yield at a similar rate to NPK mineral fertilizer	One growth season	1
Clay loamy soil. Seedlings transplanted from greenh. outdoor and planted in banks of 1 m wide x 72 m long. Urine applied at 9.7 L/m2 10, 20, 30 and 40 days after planting. Application rate: 233 kg N ha[-1] (urine) and 34 kg N ha[-1] (mineral fertilizer)	UDT[3] urine, mineral fertilizer	Cucumber[2]	The yield was not significantly different than mineral fertilizer	2-3 months	2
Field experiments on an area of 63 m[2] divided into plots of 4.5 m[2]. Fertilizer application rate - 180 kg N ha[-1]. Urine - 1.4-2.0 L/m[2]	urine, mineral fertilizer	Cabbage[3]	The yield was similar, pest resistance higher than non-fertilized control and mineral fertilizer	89 days	3
Greenhouse pot experiments of 491 cm[2]. Fertilizer application rate - 90 kg N ha[-1] (min. fert.) and 120 kg N ha[-1] (urine).	UDT Urine urine and ash, mineral fertilizer and control	Tomato[4]	Urine and ash - equal yield as mineral fertilizer and 4 times higher yield than non-fertilized plants.	88 days	4
Field plots experiments. Application rate: 50, 100, and 200 kg N ha[-1] Manual fertilizer application incorporated by hoe, one week before seedling transplant. on a	UD[6]. Faeces, goat manure and mineral fertiliser	Cabbage[5]	Higher yield, P and K crop uptake with goat manure, but lower than in mineral fertilizer	One growth season	5

18

pre-irrigated soil. Irrigation: 2 times or 3 hours per week. Cabbage seedlings transplanted to 3x3 m plots in cambisol at 36 plants per plot					
Pot experim. with 2 month old seedlings (CTR[8]) (stored for 122 days) allowed to dry (30 days) then mixed at 0, 20, 40 and 60 % (wet weight) with latosol. NPK fertilizer at 0 and 2 g per pot was disolved in 200 ml of water and mixed with latosol.	Mixed faeces and urine (CTR) Compared to mineral fertiliser and control	[7]Jatropha	The use of 40-60% of CTR incorporated into the soil significantly increased leaf number, leaf area and stem diameter, the growth being similar to mineral fertilizers or higher. Water retention in biotoilet residue was 10 times higher than in control (latosol only). Also, the N, P, K content was 35, 7 and 9 times higher than in the soil	12 weeks	6
Field experiment on volunteer farming plots Plants planted in 10x10m blocks. Fertilizer application after 4 weeks of planting at 200 ml/crop	Combined urine and faeces, control and mineral fertilizer	[9]Corn	Positive effects on corn yield and water use efficiency, higher effects compared to mineral fertilizers and non-fertilized control	One growth season	7

[1]*Ambelmoschus esculentum,* [2] control - no urine or mineral fertilizer was applied, [3]*Cucumber sativus ,* [4]*Brassica oleracea* [5]*Lycopersicum esculentum* [6]*Brassica oleracea* [7]UDT- urine diverting toilet [8], [9]CTR - composting toilet residue, [10]*Jatropha curcas* [11]*Zea mays.* References: 1 - Akpan-Idiok et. al, 2012; 2 - Heinonen-Tanski and Sjöblom, 2007; 3- Pradhan, 2007; 4 - Pradhan, 2009; 5- Mkeni and Austin, 2009; 6- Triastuti *et al.*, 2009; 7 - Guzha *et al.*, 2005.

The use of faeces or combined faeces and urine has also contributed to a significantly higher soil water retention and use efficiency in maize compared to mineral fertilizer (Guzha *et al.*, 2005). It also resulted in a higher yield of cabbage compared to goat manure (Mkeni and Austin, 2009), due to higher available phosphorus and potassium concentration. The compilation of these research findings (Table 2.1), clearly ilustrates the fertilizing value of human excreta.

2.3.2 Advantages and disadvantages of biochar application to agriculture

Application of biochar to soils with or without compost and fertilizers has been recognized among viable strategies for efficient restoration of soil organic carbon (Topoliantz *et al.*, 2005; Lehmann *et al.*, 2006; Lal, 2009). This is related to its longevity in the soil, thus its capacity to maintain stable soil organic matter for a longer period of time than application of other organic matter, e.g. compost or manure (Lehmann *et al.*, 2006). Biochars are produced by pyrolyzation of biomass in an oxygen depleted atmosphere at a high temperature of 400-800°C (Verheijen *et al.*, 2010). Hydrothermal carbonization produces another type of biochar, called hydrochar, from wet biomass at low temperature (between 180 and 300 °C) and high pressure (2 – 2.5 MPa). Application of biochar has a number of positive effects on crops and soils, including increasing the water retention and cation exchange capacity (CEC) (Abel *et al.*, 2013) and reduction of nitrous oxide (N_2O) by facilitating the transfer of electrons to soil denitrifying microorganisms (Cayuela *et al.*, 2013). As biochar contains organic and inorganic compounds which are reduced under oxygen limiting conditions, these act as electron donors under suboxic conditions for denitrifying microorganisms (Cayuela *et al.*, 2013). In paddy soils, biochar can absorb organic carbon onto its surface, thus reducing the substrate availability to methanogens and generating less CH_4 (Han *et al.*, 2016). Biochar can retain plant available nutrients, improve plant disease resistance (Elad *et al.*, 2010) and soil microbial biomass (Biederman and Harpole, 2013). It also enhances crop productivity (Liang *et al.*, 2006a; Lehmann and Joseph, 2009).

At high application rates, biochar may negatively affect plants (Buss and Mašek, 2014). Application of hydrochars or biochar produced from swine faeces caused negative effects on soil and crops, including decreases in plant-available nitrogen and reduction in plant growth (Gajić and Koch, 2012; Rillig *et al.*, 2010) and biomass compared to the control (George *et al.*, 2012) or excessive increases in soil extractable phosphorus (Novak *et al.*, 2014). One of the causes is the volatile and leachable compounds produced during pyrolysis, e.g. low molecular alcohols, ketones, phenols or polyaromatic hydrocarbons (Buss and Mašek, 2014; Lievens *et al.*, 2015). Furthermore, agricultural application of freshly produced biochar may have no or negative effects on soil and plants. This is caused by highly labile fractions of carbon that can lead to an increased mineralization of the organic matter already present in the soil (Singh and Cowie, 2014) or nitrogen immobilization and, therefore, to a reduction in crop growth and yield (DeLuca *et al.*, 2009; Rillig *et al.*, 2010). To make use of freshly produced biochars productively, co-composting and enrichment with mineral or organic fertilizers is recommended (Alburquerque *et al.*, 2013; Schmidt *et al.*, 2014). The subsequent incorporation of some aromatic fractions into the humic substances increases the humification indice of the final compost (Dias *et al.*, 2010; Jindo *et al.*, 2012). Moreover, oxidation of biochar will improve its capacity to absorb nutrients and dissolved organic matter (Cheng *et al.*, 2006; Lehmann *et al.*, 2003; Steiner *et al.*, 2010).

2.3.3 Anthrosols: land application of human excreta and biochar

The reuse of animal or human excreta in combination with biochar was used successfully in the past to transform easily weatherable sandy soils, poor in organic matter and plant available nutrients into long lasting, highly fertile lands (Pape, 1970; Lehmann *et al.*, 2003). An example is the terra preta soil type, created as a result of surface deposition, slush and burn cultivation or soil enrichment practices, with waste materials such as plant residues, human excreta and charred biomass in permanent pre-Columbian Indian settlements (Erickson, 2003; Woods, 2003). These soil types are distributed in Brazil, Columbia, southern Venezuela, Peru and the Guianas (Sombroek *et al.*, 2002; Kämpf *et al.*, 2004). Analysis of faecal specific steroids revealed a high input of human excreta into the terra preta soils (Glaser, 2007; Glaser and Birk, 2012). The biochar component contributed to soil stability and to an increased microbial biomass growth (Glaser, 2007; Glaser and Birk, 2012).

Recent studies have indicated the presence of terra preta analogous soils in Northern Europe, Australia, Asia (e.g. China and Japan) and Africa (Downie *et al.*, 2011; Frausin *et al.*, 2014; Wiedner and Glaser, 2015; Wiedner *et al.*, 2015a), where the local indigenous population had similar waste management and soil enrichment practices as those in South America. In Europe, the anthrosols were formed from heath and grass, mixed with animal manure, human faeces and charcoal there were deposited on sandy soils (Figure 2.2). The total elemental content was highly enriched compared to the reference soil, also the biochar content (30 tons ha^{-1}) was of the same magnitude as in terra preta soils (Wiedner *et al.*, 2015b).

Figure 2.2 An example of plaggen soil (grey plaggic anthrosols) from Germany (source: Giani *et al.*, 2014). The thickness of the plaggic horizon is 70-120 cm, very dark in colour and contains artefacts such as charcoal and bricks.

In China, the farmers mixed the vegetable waste with straw, turf, weeds, faeces and soil and burned it, after which they applied it to the soil (Wiedner and Glaser, 2015). From ancient times in Japan, farmers have mixed human and animal excreta with rice husk biochar and wood ash for fertilization and conditioning of the soil (Ogawa and Okimori, 2010). In Australia, the anthrosols containing charcoal were developed as a result of the disposal of waste materials from cooking to the soil. The addition of charcoal improved the soil's physical and chemical properties, including its nutrient status, cation exchange capacity and water holding capacity. In West Africa, anthrosols are still under production and use (Frausin *et al.*, 2014). The waste materials added to the soil are charcoal from cooking, residues from palm-oil soap production as well as organic materials from food preparation, crop processing, house construction and agroforestry. All these studies suggest that charcoal in combination with excreta created favourable conditions for maintaining soil fertility over long periods of time during ancient times and this practice can be applied today to restore soils around the world.

2.4 Challenges in agricultural reuse of excreta

2.4.1 Pathogen reduction

One of the main challenges for excreta application in agriculture is the risk of infecting people with disease via contaminated food crops. In human urine the number of pathogens is usually low, but potentially dangerous pathogens may still be present in endemic regions, due to urinary excretion, e.g. *Leptospira*, *Salmonella*, *Ascaris* and *Schistosoma* (Drangert, 1998; WHO, 2006). In urine-diverting toilets, cross-contamination is difficult to avoid. This also leads to an increase in the number of pathogens in urine: a study carried out in Sweden showed that 22% of the samples collected from urine tanks were contaminated with faeces (Höglund *et al.*, 2002).

In contrast to urine, human faeces contain a much higher number of bacteria, viruses, parasitic protozoa and helminths. Due to the potentially high amounts of pathogens, faeces have always been considered contagious and handled accordingly, since poor excreta sanitation may lead to the spread of disease (WHO, 2006). For example, in Vietnam, fertilization of farmlands with fresh or partially composted faecal waste from latrines caused up to 30% of hookworm infection among the local population and high indices of parasites like *Ascaris*, *Trichurus* and *Taenia* (Jensen *et al.*, 2005).

Different treatment procedures have been proposed to reduce the amount of pathogens in excreta, among which storage and composting (thermophilic composting and vermi-composting) are considered among the simplest in operation. Depending on the local climate, storage during at least 1-6 months for urine and 1-2 years for faeces is recommended for pathogen removal (WHO, 2006). Desiccation along with high pH was proven to be efficient in pathogen destruction during the storage of faeces (Niwagaba *et al.*, 2009). According to the WHO, in order to reduce the pathogens in human excreta to a safe level, a moisture content below 25% and a pH > 9 shall be ensured. In real life situations, such conditions of low moisture content and high pH can rarely be achieved. For example, wood ash with the potential to increase the pH above 9 does not always provide sufficient dehydration and thus pathogen reduction (Kaiser, 2006). Moreover, moisture levels below 25% limits the activity of decomposing bacteria and fungi (Anderson *et al.*, 1979) and therefore the faecal material during storage is more dehydrated than decomposed.

Composting is an effective method for sanitizing source-separated faeces. It does, however, need to be mixed with organic waste, like vegetable scraps, at different ratios (Koné *et al.*, 2007; Niwagaba *et al.*, 2009). A sanitizing temperature above 50 °C can be achieved and maintained for a sufficient number of days in stored faeces, if equal amounts of faeces and food waste are mixed: an *E.coli* and *Enterococcus* reduction of 3 log CFUg^{-1} and 4 log CFU g^{-1}, respectively, was achieved (Niwagaba *et al.*, 2009). Also, the addition of a carbon source, such as sawdust or straw is required to adjust the low C/N ratio of faeces to 5-10 (Gotaas, 1956). In addition, good insulation (e.g., 25-75 mm styrofoam), frequent turning and moisture levels below 65% are required.

Another way to reduce the pathogens present in faeces is via vermi-composting. The mechanisms by which earthworms contribute to the reduction of pathogens are not clearly described. Some studies have shown that the coelomocytes (a type of leukocyte) of earthworms reduce the number of pathogens, through phagocytosis, encapsulation and secreting proteins that adhere to the bacterial cells and inhibit their activities (Pablo, 2001; Popović *et al.*, 2005) during the passage of faeces through the guts of earthworms. Coliform bacteria were significantly reduced in comparison to the control only at low doses of pig slurry (approximately 1 g of slurry to 1.5 g of earthworms per day), while at a higher dose (3 g slurry : 1.5 g of earthworms per day) it did not reduce their number significantly (Monroy *et al.*, 2009). Studies also indicate that the chitinase enzyme of the earthworms can destroy the middle protective shell of helminth eggs (Hill *et al.*, 2013b), but several passages through the earthworm gut are required. However, experiments on the reduction of *Ascaris suum* egg viability did not prove efficient in comparison to the control soil without earthworms (Bowman *et al.*, 2006). Moreover, the number of viable eggs could even recover after 6 months of vermicomposting (Bowman *et al.*, 2006). An efficient reduction of *E.coli*, *Salmonella* and *Ascaris* was achieved with a high stocking density, i.e. an earthworm biomass to biosolid ratio of approximately 1 : 1.5 (Eastman *et al.*, 2001). This is as much as 40 kg of earthworms to 60 kg of biosolid, an option which is not economically feasible in all cases (Mupondi *et al.*, 2010). Vermicomposting requires an extended time of 3-5 months (Bajsa *et al.*, 2005; Sinha *et al.*, 2009), during which a significant amount of nitrogen can be lost.

2.4.2 Loss of nutrients, carbon matter and odour issues

Long term storage along with the use of alkaline covering material in faeces leads to nutrient losses via volatilization (Nordin, 2010). During extended storage and aerobic decomposition of faeces, up to 94% of the nitrogen and a significant loss of the carbon can be released to the atmosphere (Lopez Zavala *et al.*, 2002; Hotta and Funamizu, 2007). Composting also leads to nitrogen losses through volatilization, leaching and denitrification (Hao *et al.*, 2001). Up to 92% of the ammonia content can be released during composting (Eghball *et al.*, 1997), the highest ammonia volatilization being encountered during the thermophilic phase of composting, when intensive mineralization takes place (Bernal *et al.*, 1996; Hao *et al.*, 2011). Part of the nitrogen is also emitted as N_2O during the maturation phase of composting (He *et al.*, 2001). Also a significant portion of organic matter (e.g. up to 62%) is lost as CO_2 during the bio-oxidative stage of composting (Bernal *et al.*, 1996). During vermi-composting, most of the nitrogen is emitted as N_2O due to incomplete denitrification processes occurring in the guts of the earthworms (Frederickson and Howell, 2003). The variations in the nitrogen and organic matter loss depend on the time required for decomposition of the organic waste.

Urine storage also leads to nutrient loss. Under the influence of urease positive bacteria and free urease, urea is degraded to ammonia (NH_4^+) (Udert *et al.*, 2003; Udert *et al.*, 2006), part of which can be lost during storage, transportation and field application (Sherlock and Goh, 1984; Rodhe *et al.*, 2004). The release of ammoniacal nitrogen increases at pH values above 8 (Williams *et al.*, 2011). Urine hydrolysis also facilitates the precipitation of phosphates and carbonates, which are deposited at the bottom of the collection tanks (Udert *et al.*, 2003). Repeated application of stored urine to the soil can cause salinization, due to its ionic composition (Ca^{2+}, Mg^{2+} and SO_4^{2-}) and a major source of soluble salts as NaCl (Neina and Dowuona, 2014). Ammonia and other malodourous compounds (e.g. VFA, indolic and phenolic compounds) formed during organic decomposition cause an undesirable odour of the stored urine (Troccaz *et al.*, 2013; Zhang *et al.*, 2013), which might be of considerable concern during agricultural applications.

Chemical acidification prevents volatilization of nitrogen, thus increasing the fertilizing value of excreta. For example, acidified cattle slurry had a 26 % higher nitrogen content than the non-acidified one (Kosmalska, 2012). Besides reducing the NH_3 loss, acidification of the slurry substantially reduces CH_4 emissions (Petersen *et al.*, 2012). However, this method is costly compared to the values of increased nitrogen retention (Kirchmann, 1994). Acidification with sulphuric acid can cause H_2S emissions and may negatively impact the soil by raising its electric conductivity (Frost *et al.*, 1990). In addition, acid treatment may cause an increase in the concentrations of volatile fatty acids and volatile sulphurous compounds that may add to an overall increase in odour emissions (Ottosen *et al.*, 2009; Petersen *et al.*, 2012). Therefore, lowering the pH value can be a promising possibility for reducing gaseous emissions in excreta. However, chemical acidification may be undesirable due to potential environmental concerns and odour intensification. Alternative possibilities that use environmentally friendly bio-based products shall be considered.

2.5 Lactic acid fermentation of human excreta

2.5.1 Transformations occurring during lactic acid fermentation

Lactic acid fermentation has been widely applied in food and silage preservation, treatment of kitchen and agricultural waste as well as animal manure (Kamra *et al.*, 1984; Andersson *et al.*, 1988; Wang *et al.*, 2001; Schroeder, 2004; Murphy *et al.*, 2007). The use of lactic acid fermentation for treatment of human excreta has been extensively studied within the terra preta sanitation approach (Factura *et al.*, 2010; Otterpohl and Buzie, 2011; Yemaneh *et al.*, 2012; Otterpohl and Buzie, 2013; Schuetze and Santiago-Fandino, 2014; Yemaneh *et al.*, 2014; Bettendorf *et al.*, 2015).

Lactic acid fermentation consists of two major stages: an aerobic and an anaerobic one. During the first stage, oxygen is consumed by a mixed culture of aerobic microorganisms (Woolford, 1984). During the second stage, the soluble carbohydrates are transformed to organic acids, ethanol, acetaldehyde and carbon dioxide by homofermentative and heterofermentative lactic acid bacteria (LAB) (Murphy *et al.*, 2007). The accumulation of high amounts of organic acids results in a pH decline to a level that inhibits further microbial growth. The lacto-fermented material thus becomes stabilized and can be preserved for a long time.

The extracellular enzymes produced by fermentative bacteria contribute to the breakdown of the polysaccharides, proteins and lipids into smaller monomers, which are further used for synthesis of organic acids and microbial exopolysaccharides or for bacterial growth (De Vuyst and Degeest, 1999; Sánchez, 2009). LAB produce a wide range of enzymes such as hydrolytic, carbohydrate degrading enzymes, proteolytic, lypolytic and starch-modifying enzymes (Patel *et al.*, 2013; Petrova *et al.*, 2013), which solubilise the solid substrate. Thus, a better substrate utilization by bacterial and fungal consortia is possible when additional treatment, such as composting, is performed after the lactic acid fermentation (Xavier and Lonsane, 1994; Yang *et al.*, 2006). As the excreted faeces contain cellulolytic bacteria such as *Clostridium* and *Bacteroides*, also non-digestible compounds like lignocellulotic materials (e.g. sawdust from cover material or toilet paper) can be partly decomposed during lactic acid fermentation (Bryant, 1978; Wedekind *et al.*, 1988; Kalantzopoulos, 1997).

With a sufficient amount of organic acids, there is a pH decrease, which prevents the degradation of proteins; does not significantly change the nutrient contents; and lowers the organic matter loss. For example, the protein content did not change significantly after 10 days of lactic acid fermentation of poultry manure with 10 % of molasses and 3 % of LAB inoculum at pH 4 (El-Jalil *et al.*, 2008). In another study on a corn-broiler litter mixture, the total nitrogen content as well as crude protein content changed insignificantly even after 80 days of ensiling (Caswell *et al.*, 1977). Also, treatment of manure with acidic liquid bio-waste from the milk and citrus industry decreased the pH to 5, reducing the CH_4, N_2O and NH_3 emissions by 76-78%, 36-37 % and 84-86 %, respectively (Clemens *et al.*, 2002; Samer *et al.*, 2014). After treatment of manure with lactic acid, no N_2O emission was recorded anymore and a 75-90% reduction of CH_4 emission was observed (Berg *et al.*, 2006).

Lactic acid fermentation can contribute to the reduction of pathogens in excreta (Scheinemann et al., 2015; Yemaneh, 2015). However, its efficiency in the destruction of resistant pathogens is still insufficiently investigated. An effective pathogen reduction was attained with the addition of molasses and wheat bran as source of carbohydrates, which contributed to the formation of lactic acid and pH reduction. Lactic acid fermentation of faecal sludge and manure contributed to effective pathogen reduction with the addition of molasses and wheat bran (10-12 %) as sources of soluble carbohydrates (McCaskey and Wang, 1985; Scheinemann *et al.*, 2015). Lactic acid fermentation of swine manure has decreased the concentration of the *Enterobacteriacea*, *Staphylococcus* and *Clostridium* populations by 7 log $CFUg^{-1}$ after 5 - 10 days, when the pH was reduced to 4 (El-Jalil *et al.*, 2008). In another study on swine manure with corn, a complete destruction of faecal coliforms, *Salmonella derby* and *Treponema hyodysenteriae* was obtained during 3 weeks of ensilage at pH 4 (Weiner, 1984). The pathogeneity of *Ascaris suum* was lost completely after 56 days its viability was already reduced after 21 days (Scheinemann *et al.*, 2015) at pH 5.0-5.2. In a study on swine manure, *Ascaris suum* eggs remained infective even after 56 days of fermentation processing at pH 3.9-4.4, even though their viability was considerably reduced (Caballero-Hernández *et al.*, 2004). Further investigations are needed on the efficacy of combined lactic acid fermentation and thermophilic composting or vermi-composting in the treatment of faeces.

The hygienization of faeces by lactic acid fermentation is caused by pH reduction and the production of lactic acid and other compounds with antagonistic activities to fungi, protozoa and a wide range of gram positive and gram negative bacteria (Rattanachaikunsopon and Phumkhachorn, 2010). Protein complexes like bacteriocins and other compounds, e.g. glucose oxidase, hydrogen peroxide and exopolysaccharides have also a suppressive effect on pathogenic microorganisms (Patel *et al.*, 2012; Saranraj, 2014). The mechanisms of the reduction of helminths eggs are poorly described. One of the factors is the increased temperature (36-37 °C) during fermentation (Scheinemann *et al.*, 2015). As lactic acid fermentation does not produce a temperature increase and the process usually takes place at lower temperatures than those indicated above, it might be challenging to achieve this.

Lactic acid fermentation also deodorizes the offensive odour of excreta, as LAB inhibit microorganisms producing malodourous ingredients, e.g. S-compounds (H_2S), N-compounds (NH_3, indole and scatole) and C-compounds (lower fatty acids) (Hata, 1982; Zhu, 2000; Wang *et al.*, 2001). Formation of volatile fatty acids, responsible for the odour production in faeces, is inhibited during the lactic acid fermentation process (Kamra *et al.*, 1984; Yemaneh *et al.*, 2014). After a few days of fermentation, the foul faecal odour is changed to a sour-silage-like one. Urine lacto-fermentation leads to odour reduction by a decrease in malodourous compounds (Yemaneh *et al.*, 2012; Zhang *et al.*, 2013). For example, it was found that a range of urease positive bacteria such as *Escherichia fergusonii*, *Enterococcus faecalis*, *Citrobacter kaseri*, *Streptococcus agalacticae* and *Morganella morganii* are responsible for the representative stale urine odour due to emission of phenol, indol and sulphide (Troccaz *et al.*, 2013). LAB inhibit the activities of these organisms by producing a wide range of secondary metabolites (Savadogo *et al.*, 2004).

Considering the potential of lactic acid fermentation for hygienization as well as reduction of nutrient loss and odour emissions, this treatment technique can be applied in mixed and separately collecting sanitation systems (Windberg *et al.*, 2013; Anderson *et al.*, 2015). In UDDTs, the addition of LAB and accompanying sugar-containing substrates can be done both at the collection and post-collection stages (Figure 2.3).

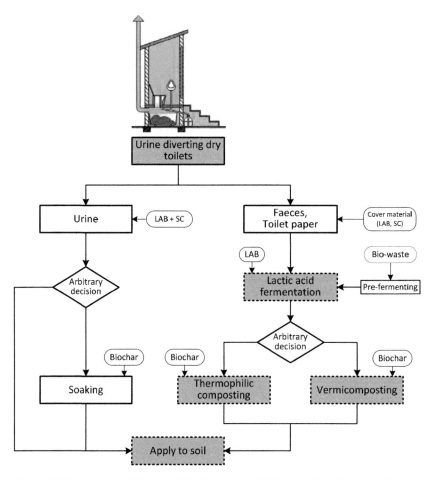

Figure 2.3 Overview of the potential steps and additives used during lacto-fermentation of the urine and faeces fraction of urine diverting dry toilet systems
LAB: lactic acid bacteria, SC: soluble carbohydrates.

2.5.2 Lactic acid fermentation of urine

At the collection stage, urine can be treated with a solution of LAB and soluble carbohydrates, e.g. molasses or whey (Figure 2.3). A limited number of studies (Reckin, 2010; Beriso and Otterpohl, 2013; Wolf, 2013) report about urine treatment via lactic acid fermentation. Furthermore, there is only scarce information regarding the added agricultural or hygienic value. Some medical studies indicate that LAB strains such as *Lactobacillus casei*, *L. acidophilus*, *L. gasseri* and *L. plantarum* produce substances that inhibit the activity of urease positive bacteria, for example in the pathogenic bacterium *Helicobacter pylori* (Michetti *et al.*, 1999; Servin, 2004; Sgouras *et al.*, 2004). Among the potential inhibitory mechanisms, the production of bacteriocins is mentioned (Tabak *et al.*, 2012). Similarly to this, during lactic acid fermentation of urine, LAB can inhibit urease positive bacteria, thus preventing urea hydrolysis.

During storage, the urine pH increases rapidly up to 8-9, when most urea is hydrolyzed (Kirchmann and Pettersson, 1994; Udert *et al.*, 2003). The hydrolyzed urine has a high buffering capacity, it would be more appropriate if the soluble carbohydrate source and the LAB are added to storage tanks prior to urine starts to accumulate there. Non-hydrolyzed urine could be a good growth medium for LAB as it contain urea, amino acids and minerals such as K^+, Na^+, Mg^{2+}, PO_4^{3-}, SO_4^{2-} and Cl^- which are important for the growth of LAB (MacLeod and Snell, 1947; Udert *et al.*, 2006). Experiments show good acidification effects of the sauerkraut juice microbial inoculum to the fresh urine with a ratio of 1:6 by decreasing the pH to 4 after 1.5 month treatment (Wernli, 2014). During the urine lactic acid fermentation, the LAB inhibit the bacterial urease, thus preventing urea hydrolysis and consequently reduce ammonia volatilization. For example, studies have shown that the urease activity is diminished at pH < 5 (Larson and Kallio, 1954; Schneider and Kaltwasser, 1984). The highest release of NH_3 takes place between pH 7 and 10; below pH 7, NH_3 volatilization decreases and at pH 4.5 free ammonia is completely unavailable (Hartung and Phillips, 1994; Williams *et al.*, 2011) (Hartung and Phillips, 1994; Williams *et al.*, 2011).

Further research on the application of lactic acid fermentation of urine is important considering the potential for reducing the pH, thus minimizing the ammonia loss and phosphorous precipitation as well as the associated consequences such as smell and pipe clogging. It is also important to investigate the effects of lactic bacteria on malodourous components in urine. Lactic acid fermentation can further add to a reduction of pathogens that may result from faecal cross-contamination. The capacities of LAB to degrade micropollutants such as hormones, pharmaceuticals and pesticides which may be excreted via urine and diffuse into the aquatic environment after urine application in agriculture need further investigation. Ensilage of corn grain mixed with broiler litter for a period of 80 days reduced significantly ($p < 0.01$) the concentration of sulfoquinoxaline, a veterinary medicine to treat coccidiosis compared to before ensilage (Caswell *et al.*, 1977). Some LAB strains are effective in the decomposition of organophosphorus pesticides, using them as a source of carbon and phosphorus for the synthesis of the enzyme phosphatase (Cho *et al.*, 2009; Zhang *et al.*, 2014b).

Lacto-fermented urine can also be soaked on biochar for nutrient capturing. Application of wood biochar could reduce ammonia volatilization from ruminant urine by 45 %, while retaining its bioavailability to plants (Taghizadeh-Toosi *et al.*, 2011). Acidified urine may be more effective for biochar charging because sorption of ammonia and phosphate onto biochar is higher at lower pH (Yao *et al.*, 2011; Spokas *et al.*, 2012). For example, phosphate adsorption was maximal at pH 2 to 4.1 and decreased at pH above 6 (Yao *et al.*, 2011). The ions and organic molecules in the urine can enter the pore system of the biochar particles (Schmidt *et al.*, 2015). Besides, the organic compounds in the urine are adsorbed on the biochar surface, forming an organic coating onto which anions and cations from urine (e.g. phosphates and ammonium) are bound (Schmidt *et al.*, 2015).

2.5.3 Lactic acid fermentation of faeces

Fresh human faeces contain a low number of microbial species facilitating lactic acid fermentation, e.g. *Lactobacillus, Pediococcus* and *Leuconostoc* (Sghir *et al.*, 2000; Dal Bello *et al.*, 2003). Thus, the addition of lactic acid-producing bacteria as well as substrates favouring their growth is required (Otterpohl and Buzie, 2013). The use of lactobacilli strains such as *Lactobacillus plantarum*, *Lactobacillus casei* and *Pediococcus acidilactici* as additives promotes effective lactic acid fermentation of faeces (Yemaneh *et al.*, 2012). Low cost sources of LAB are waste products of milk, fish, meat and grains (Frazier, 1967; Park and DuPonte, 2008; Windberg *et al.*, 2013; Bulbo *et al.*, 2014). The outer layers of grain, especially rice, contain essential nutrients, carbohydrates, vitamins as well as microorganisms from the families of *Lactobacilacea, Pseudomonacea, Micrococcacea,* and *Bacilacea* (Frazier, 1967; Saman *et al.*, 2011). Therefore, rice wash water can serve both as a growth medium and inoculum.

The excreted carbohydrate fraction in faeces, which is an important substrate for lactic acid bacteria, is usually composed of undigested cellulose, vegetable fibre, oligosaccharides and polysaccharides, not hydrolyzed by the intestinal secretions of humans (Canfield and Goldner, 1964; Southgate and Durnin, 1970). Soluble carbohydrates can be found in molasses, whey or waste from starchy materials such as potato, wheat, manioc or barley (Anuradha *et al.*, 1999; Pandey *et al.*, 2000). A potential cheap source of soluble carbohydrates that can be added to the cover material for faeces dehydration is press mud, a non-value waste from the sugar industry which contains 5-15 % sugar (Xavier and Lonsane, 1994; Solaimalai *et al.*, 2001; Partha and Sivasubramanian, 2006) and is produced in large amounts. For example, the sugar industry in India produces annually 5.2 million tonnes of press mud (Rakkyappan *et al.*2001). Studies have shown that this material was successfully used as a mixing agent in silages (van der Poel *et al.*, 1998). Press mud addition is important if lactic acid fermentation is followed by vermi-composting, as it is easily processed by earthworms (Prakash and Karmegam, 2010). Additional components in the covering material can be chopped corn stover, clay and rock flour (Snyman *et al.*, 1986; Bottcher *et al.*, 2010; Bagar and Kavčič, 2013). Grinded corn stover was efficiently lacto-fermented with cattle manure (Murphy *et al.*, 2007). The rock flour stimulates microorganisms by providing them a habitat and the necessary micronutrients required for the breakdown of organic matter. Clay and rock flour will add to the formation of organo-mineral chelated complexes, thus increasing organic matter stability (Stevenson, 1994; Bagar and Kavčič, 2013).

Some authors indicate that a mixture of ground biochar can also be added to the covering material at the collection stage of faeces (Factura *et al.*, 2010; Windberg *et al.*, 2013). The application of biochar as a matrix in composting toilets showed a 2.3 and 1.7 times lower carbon loss and a 1.9 and 1.3 lower nitrogen loss than rice husk and corn stalk, respectively (Hijikata *et al.*, 2015). Since the pH of the biochar is mostly alkaline (Verheijen *et al.*, 2010) it may hinder the fermentation process. Therefore, its addition at the end of the lactic acid fermentation (Yemaneh *et al.*, 2014) or at the post-treatment stage (thermophilic or vermi-composting) is more appropriate (Bettendorf *et al.*, 2014).

The quantity of source separated faeces and toilet paper generated per capita per year is relatively low, approximately 70 kg and 6 kg (wet weight) per year, respectively (Rose *et al.*, 2015). After the collection stage, faeces and toilet paper can be treated via extended lacto-fermentation along with bio-waste such as kitchen/food industrial waste (e.g. fruit, milk or sugar industry wastes) or animal manure (Figure 2.3). Vegetable waste is rich in LAB, enzymes (e.g. cellulolytic, lignolytic and pectinolytic) and carbohydrates (Jawad *et al.*, 2013) and can increase the efficiency of lacto-fermentation of faeces and cover material as well as of the thermophilic stage of composting, thus shortening the overall time required for the treatment of faeces (Ong *et al.*, 2001). Pre-fermenting (simple storage in closed vessels) of kitchen and fruit waste prior to being mixed with the faeces will release the sugars under the influence of extracellular enzymes and facilitate a better fermentation of faeces. Addition of cattle or swine manure to faeces during secondary faeces treatment is beneficial owing to a higher number of lactic, cellulolytic and hemicellulolytic bacteria, that are capable of fermenting a range of monosaccharides, oligosaccharides, polysaccharides and lignocelluloses (Cotta *et al.*, 2003; Dowd *et al.*, 2008). These cellulolytic and hemicellulolytic bacteria, e.g. *Clostridia* and *Ruminococcus* (Sijpesteijn, 1951; Wang *et al.*, 2007; Dowd *et al.*, 2008), are important for an effective decomposition of lignine and cellulose from toilet paper and sawdust in the cover material, before soil application. Micro-aerobic conditions occur in compost when the oxygen concentration is less than 2.0% throughout the whole composting process (Wang *et al.*, 2001). Under these conditions, lignocellulose degradation takes place at the initial composting stage as the bacteria which decompose celluloses are more effective.

2.5.4 Limitations of lactic acid fermentation

Among the main limitations of lactic acid fermentation is the need for creating an enabling environment for the optimal growth of LAB, promoting acidification and stabilization of excreta, e.g. provision of LAB strains and nutritional additives like a carbon source (Saeed and Salam, 2013). The nature of added components (e.g. kitchen, waste from food industry or cattle manure) as well as the interactions between different LAB strains and fungi during the storage and the fermentation process may also influence the efficiency of the lactic acid fermentation or combined lactic acid fermentation and composting (aerobic thermophilic composting or vermi-composting) processes. These affect the structure of the substrate, oxygen permeability, moisture content and pH value.

Another limitation is that an effective acidification of faeces is obtained only with the addition of sources of easily fermentable carbohydrates such as molasses and wheat bran (Yemaneh *et al.*, 2012; Scheinemann *et al.*, 2013; Böttger *et al.*, 2014). These may have other potential competitive uses, such as in the fermentation industry, production of renewable fuels or food and feed additives (Leng, 1984; Noike and Mizuno, 2000; Dumbrepatil *et al.*, 2008; Siqueira *et al.*, 2008; Javed *et al.*, 2012). Therefore, it is important for revealing non-value waste materials rich in sugar or starch, such as kitchen waste and food industry waste, to be used in lactic acid fermentation (Samer *et al.*, 2014; Yemaneh *et al.*, 2014).

2.6 Post treatment of lacto-fermented human excreta

Lactic acid fermentation of faeces contributes to hygienization, pre-decomposition of the digestate and odour reduction. However, direct application of lacto-feremented faeces to the soil may create some problems. There are a number of constrains for soil application (Figure 2.4): the lacto-fermented faeces may not always be fully sanitized (Factura *et al.*,

2010; Buzie and Körner, 2014), the lacto-fermented faeces is anaerobic and rich in organic acids, its addition to the soil may cause incomplete denitrification or intensification of the mineralization of soil organic matter (Hamer and Marschner, 2005; Green and Popa, 2011).

In poorly drained soils, an accumulation of organic acids may occur with a phyto-toxic effect on plant roots (Fu, 1989). As the lacto-fermented material may be incompletely mineralized, its direct application to the soil may cause a rapid decomposition, leading to a decrease in oxygen concentration in the root zone and potentially to an increase in the solubility of heavy metals (Jiménez and Garcia, 1989). Therefore, before being applied to the soil, the lacto-fermented faeces can be either mixed with biochar or processed further via vermi-composting or thermophilic composting.

2.6.1 Vermi-composting

Vermi-composted material has a better structure as the earthworms are mixing and grinding the substrate, thus increasing nutrient availability and the content of humic substances (Figure 2.3). The mineralization activity of bacteria, fungi, and actinomycetes populating the earthworm gut release plant available nutrients, for example N, P K and trace minerals (Zn, Mn and Fe) (Yadav and Garg, 2011). One limitation of this treatment option is that it needs to be pre-conditioned, e.g. ventilated for at least 72 hours or mixed with bulking material such as vermi-compost, cattle manure or biochar prior to inoculation with earthworms (Buzie and Körner, 2014). Limited oxygen supply in the lacto-fermermented material and low pH may be detrimental to earthworms as they die very quickly in anaerobic conditions (Munroe, 2007). The lacto-fermented mix is also rich in organics (lactic and acetic

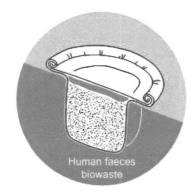

LACTO-FERMENTATION

Effects on faeces:
hygienization, pre-decomposition, odor
reduction

Constraints for soil application:
insufficient hygienization, anaerobic,
incomplete mineralization

VERMICOMPOSTING

- Supplementary hygienization
- Nutrient availability, humification
- Mixing/grinding

THERMOPHILIC COMPOSTING

- Supplementary hygienization
- Nutrient availability, humification
- Elimination of weeds

Figure 2.4 Effects of lacto-fermentation on human faeces, constraints for soil application and post-treatment via combined lacto-fermentation with vermi-composting or thermophilic composting. *Human faeces mixed with bio-waste can be lacto-fermented under conditions of low oxygen supply (e.g. pit earths covered with plastic foil) during which pathogens and odour are reduced. Due to incomplete decomposition and insufficient pathogen removal, additional treatment of excreta needs to be performed via a composting stage (this could be either themophilic or vermi-composting.*

acids) and has a low pH that might have a deleterious effect on earthworms (Frederickson and Knight, 1988). For example, earthworms introduced in lacto-fermented faeces died within 24 hours (Factura *et al.*, 2010).

Vermi-composting and earthworm activities generate high N_2O emissions from the worms themselves and their casts (Horn *et al.*, 2003). Treatment of faeces and bio-waste via lactic acid fermentation results in a high content of organic acids and amino acids, which decrease the pH. The sugar and organic acids can serve as electron donors for nitrate reduction in the gut of earthworms and thus can lead to more intense N_2O emissions in the course of incomplete denitrification (Karsten and Drake, 1997; Asuming-Brempong and Nyalemegbe, 2014). Combining lactic acid fermentation with thermophilic composting is thus a more preferred option than lactic acid fermentation with vermi-composting.

2.6.2 Thermophilic composting

Another way to improve the structure of the lacto-fermented faeces is through composting. Compost maturation can be achieved with passive aeration and limited turning only (micro-aerobic composting), with an oxygen supply of less than 2 % or by turning and forced aeration (aerobic composting) (Wang *et al.*, 2001). For example, the soluble sugar content was significantly lower, while the rate of cellulose degradation was significantly higher after 15 days of composting under micro-aerobic composting compared to aerobic composting. Furthermore, based on the C:N ratio, the compost was mature at 45 days under micro-aerobic conditions and at 60 days under aerobic conditions (Wang *et al.*, 2001). Micro-aerobic composting with limited oxygen supply might be more efficient for the degradation of lacto-fermented substrates. The presence of labile forms of carbon in the initial material (caused for example, by the addition of molasses) contributes to a reduction of ammonia volatilization due to immobilization of the NH_4^+-N by the microbial biomass. Provision of extensive air supply and turning is counterproductive as it leads to an intense release of nitrogen (NH_3-N) from the composting material (Liang *et al.*, 2006b).

During the thermophilic phase of micro-aerobic composting, along with labile carbon, slower degradable materials such as celluloses and hemicelluloses are also degraded by facultative thermophile anaerobes, e.g., *Clostridia*, whose cellulolytic enzyme system works optimally in micro-aerobic conditions, with O_2 concentrations below 2 % and temperatures of 50-60 °C (Schwarz, 2001; Wang *et al.*, 2007). In contrast, lignocellulolytic degradation will be inhibited in aerobic composting and fungi and actinomycetes will perform degradation of these materials at the end of the composting period. Therefore, the aerobic composting process is taking longer to stabilize, it also requires turning and aeration (Wang *et al.*, 2007). With shorter turnover rates, micro-aerobic composting can recycle carbon matter to the soil more efficiently than aerobic composting or vermi-composting. For example, during micro-aerobic composting only 2% of carbon is lost as CO_2, compared to 41% and 63%, respectively, during aerobic composting and vermi-composting (Green and Popa, 2011; Fornes *et al.*, 2012).

2.6.3 Addition of biochar

Biochar has a high surface area (Jindo *et al.*, 2012) and adsorbs NH_3, NH_4^+ and NO_3^-, thus preventing their loss via volatilization or leaching. By creating a favourable microenvironment for nitrifying bacteria (Chowdhury *et al.*, 2014; Jindo *et al.*, 2016; López-Cano *et al.*, 2016), it also promotes nitrification. Biochar particles further improve the texture of the lacto-fermented material thus improving aeration (Jindo *et al.*, 2012). Biochar addition at the vermi-composting stage is important for increasing the stability of the compost, reducing the nutrient loss and improving the compost structure. Earthworms ingest biochar particles, grind and excrete them in the vermicasts, thus mixing them with the compost (Asuming-Brempong and Nyalemegbe, 2014; Eckmeier *et al.*, 2007).

Adding biochar during micro-aerobic composting will improve the reduction of pathogens in faecal matter by increasing the peak thermophilic temperature and shortening the time for reaching the sanitizing temperature above 55 °C (Steiner *et al.*, 2010). The high temperature (± 70°C) generated by aerobic thermophilic composting further adds to surface oxidation and an increase in the acidic functional groups, thus enhancing the biochar capacity to absorb nutrients and dissolved organic matter (Cheng *et al.*, 2006; Dias *et al.*, 2010; Jindo *et al.*, 2012). It also accelerates humus formation due to the incorporation of biochar aromatic fractions into the humic substances of the composted material (Dias *et al.*, 2010; López-Cano *et al.*, 2016; Steiner *et al.*, 2010). A high degree of humus formation was observed upon the addition of only 2% dry weight of biochar to compost (Jindo *et al.*, 2012).

2.7 Agricultural effects of lacto-fermented excreta

2.7.1 The effects of lacto-fermented excreta on soil and plants

Research on the agricultural effects of lacto-fermented human excreta is rather scarce. The benefits reported are mainly traced from related studies on bokashi (anaerobically fermented animal manure or vegetable waste with rice bran and microbial inocula) and also on LAB. Studies have evaluated the improved growth, yield and quality of crops, increased nutrient availability (e.g. phosphorus), enhanced soil biological activity and physical soil properties as well as suppression of plant pathogens. Table 2.2 presents an overview of the main effects on soils and crops of lacto-fermented human and animal excreta and vegetable waste.

Application of lacto-fermented faeces and biochar to mung bean (25:75 residue : clay soil ratio) contributed to a higher plant growth compared to the unfertilized control (Prabhu *et al.*, 2014). In a two year field experiment, lacto-fermented faeces and biowaste, supplemented by urine charged biochar, significantly improved (p < 0.05) the yield of corn on a clay-loamy chernoziom in Moldova compared to the unfertilized control, stored cattle manure, faeces, urine and vermicomposted lacto-fermented mix. In contrast, the mineral fertilizer gave a significantly lower corn yield during the first production year, but not significantly different yield during the second production year. This fertilizing mix has also significantly reduced the soil bulk density and increased the mobile potassium content in comparison to the control and other fertilizers (Andreev *et al.*, 2016).

Table 2.2 Overview of the main effects on soil and plants of the application of lacto-fermented faeces, lacto-fermented and composted organic waste, and biochar

Experimental conditions	Soil type, geographical location and scale of the experiment	Plants used	Effects on soil and crops	Reference
Sewage sludge, charcoal (20%), and lactobacilli (2%) were lacto-fermented for 28 days in enclosed buckets, followed by vermin-composting with *Eudrillus eugenia* earthworms.	Acidic clay soil (pH=5.5-6.5), low in P and Mg. Goa, India Pot experiments	Mung bean (*Vigna radiata*)	Plants with 25 % lacto-fermented sludge had the highest leaf length and leaf area compared to other treatments.	Prabhu *et al.*, 2014 a, b
Source-separated feces and biowaste (manure, kitchen waste) was lacto-fermented with sugar beet molasses and bacterial inoculum from sauerkraut in pits in the ground lined and covered with plastic for 8 weeks. Before field application, biochar that had been soaked in urine was mixed in. For the second production year, this mix was also vermicomposted.	Clay loamy chernoziom Eastern Europe, Moldova Two-year field experiments	Corn (*Zea mays*)	Yield was significantly higher than the control, stored cattle manure, feces, urine and vermicomposted lacto-fermented mix. In relation to mineral fertilizer, a significantly higher yield was produced during first year but not the second. Lacto-fermented mix with biochar had lowered soil bulk density and enhanced soil potassium content in relation to the control and other fertilizers.	Andreev *et al.*, 2016
Bokashi [1] and mixed culture of LAB, yiest and photosynthetic bacteria - EM[2] compared to NP[3] fertilizer. Different treatments included: drainage and without drainage; bokashi; and	Field experiments China, loam clay saline soil	Rice (*Oryza sativa*)	EM bokashi and subdrainage had significantly lowered bulk density, increased organic matter content, available P, cation exhange capacity and microbial biomass compared to all other treatments. Also EM	Xiaohou *et al.*, 2008

mineral fertilizers			bokashi and drainage has significantly increased the yield and quality of rice.	
Bokashi (Rice bran, rice husk, rapeseed oil mill sludge, fish meal anaerobically fermented with *Lactobacillus* and yeast), application rate 3000 kg/ha compared to NPK fertilizer (15:15:15) at a rate of 600 kg/ha	Field experiments 42 m2 each experimental plot	Peanut, (*Arachis hypogaea*)	Fresh weight of nodules on lateral plants, the number of pods with two seeds, and yield in plants fertilized by EM bokashi was significantly higher than in those fertilized by chemical fertilizer	Pei-Sheng and Hui-Lian, 2002
Bokashi was obtained by fermenting fruit waste, grass weeds with molasses and effective microorganisms in a closed bucket for a period of 10 days. Comparison was made with bokashi without EM and control of farmer's normal practice.	Field experiments on a 197.4 m^2 area	Corn (*Zea mays*)	Plants fertilized with EM bokashi were two times taller and had 14% more leaves	Alattar *et al.*, 2016

[1] rice bran and cattle manure (90.4%), water (9.0%) with molasses and effective microorganisms
[2] EM - mixed culture of lactic acid bacteria, yeast and photosynthetic bacteria (0.6%).
[3] nitrogen and phosphorus fertilizer

Fermented cattle manure with molasses and effective microorganisms (mixed culture of LAB, yeast and photosynthetic bacteria) has a decreased bulk density, but an increased organic matter content, available phosphorous, cation exchange capacity, microbial biomass as well as yield and quality of rice compared to mineral fertilizer (Xiaohou *et al.*, 2008). Anaerobically fermented rice bran, rice husk, rapeseed oil mill sludge and fish meal fermented with *Lactobacillus* and yeast has significantly increased the fresh weight of nodules, pod numbers and yield of peanut compared to chemical fertilizer (Alattar *et al.*, 2016). The application of lacto-fermented kitchen scraps produced twice taller corn plants (p

< 0.05) and more leaves than the control (Alattar *et al.*, 2016). Fermented organic waste with wheat bran and effective microorganisms improved nutrient availability, but produced a 29 % lower tomato yield than mineral fertilizer mixed with effective microorganisms (Hui-Lian *et al.*, 2001). The quality of the fruits, e.g. sugar, organic acid and vitamin C concentration was, however, higher in plants amended with fermented organic waste.

The addition of lacto-fermented material affects different biological, physical and chemical soil components caused by the LAB themselves as well as by the compounds they produce. The application of lacto-fermented vegetable waste and effective microorganisms led to a 35 %, 23 % and 12 % increase of soluble N, P and K, respectively, compared to the unfertilized control (Lim *et al.*, 1999). LAB are able to solubilize water insoluble phosphates (Zlotnikov *et al.*, 2013) from compost and also from the soil, through mechanisms similar in phosphorus solubilizing bacteria, e.g. production of organic acids, pH lowering and phosphatase enzymes (Park *et al.*, 2010), thus increasing its availability to plants. In addition, some of the organic acids (e.g. lactic or acetic acid) produced by LAB can be used as carbon source by other useful microorganisms, such as photosynthetic bacteria, able to fix N_2 (Kantha *et al.*, 2012).

LAB can beneficially influence the growth and yield of crops. For example, tomato amended with LAB had 2-4 fold more fresh weight of fruits than the unamended control plants (Hoda *et al.*, 2011). *Lactococcus lactis* showed significantly higher growth than the control, in cabbage grown under greenhouse conditions (Somers *et al.*, 2007). Different compounds such as carbohydrates, amino acids and organic acids and other metabolites produced by the LAB may stimulate soil beneficial microorganisms and suppress phyto-pathogens (Hoda *et al.*, 2011). In a pot experiment, LAB have suppressed by 60 % the growth of plant pathogen *Phytium* (Hoda *et al.*, 2011). LAB have reduced by 63 % the disease incidence of bacteria wilt *Ralstonia solanacerum* in tomato (Murthy *et al.*, 2012) and have caused significantly fewer damages upon a bacterial infection by *Pseudomonas* (number of lesions per leaf and percentage of dead leafs; $p < 0.05$) in beans compared to the control (Visser *et al.*, 1986).

2.7.2 Effect of post-treatment by composting and biochar addition

Compost obtained from combined lactic acid fermentation and vermi-composting is enriched with humic acids and phytohormones (Arancon *et al.*, 2010; Zhang *et al.*, 2014a). Moreover, it has improved physical characteristics such as porosity and water holding capacity. The quality of the obtained faecal compost treated via combined lactic acid fermentation and vermi-composting had a higher total nitrogen, phosphorus and total organic carbon content than certified compost (Bettendorf *et al.*, 2014). A pathogenically safe compost was obtained from combined lactic acid fermentation and thermophilic composting, which beneficially influenced the germination of radish. Therefore, seeds treated with this type of compost had a germination index of 90 % of the control, which was distilled water (Andreev *et al.*, 2015).

The biochar component may also provide beneficial effects on different soil characteristics such as the structure, texture, porosity, particle size distribution, density, oxygen content, water storage capacity, microbial and nutrient status in the root zone (Atkinson *et al.*, 2010). The main value of mixing biochar with excreta is the reduction of nitrogen loss. NO_3^- is captured and retained on the biochar during the composting process, thus will not be readily leached which makes it available for plants. For example, in a lysimeter study with pasture soil, pine forestry plantation soil and soil amended with biosolids in combination with biochar have contributed to a decrease in nitrate leaching compared to the control soil (no biochar, no biosolids) and soil amended with biosolids only (Knowles *et al.*, 2011). Pot experiments with soil and gravel and addition of composted faeces from compost toilets, where biochar was used as a matrix during the collection stage, increased significantly the growth of mustard spinach *Brassica rapa* compared to sandy soil, without nutrient addition (Hijikata *et al.*, 2015). The authors related this growth increase to the biochar contribution to NO_3^- retention, leading to increased nitrification rates as well as preventing its loss via leaching. Biochar can absorb NO_3^- onto its surface, however, nitrification could not be measured in this study (Clough *et al.*, 2013). The fact that biochar adds with improved aeration can reduce the activity of denitrifying bacteria which act under anoxic conditions, thus preventing NO_3^- loss via denitrification (Wang *et al.*, 2013; Zhang *et al.*, 2010).

Compost supplemented by biochar (2% wet weight) improved up to 5 times the growth of *Chenopodium quinoa* (Kammann *et al.*, 2015). In other studies, for example in pot experiments conducted with a ferrasol with biochar amended compost and biochar as a substitute for peat growth medium, biochar contributed to increased water retention, nutrient uptake by plants and crop yield (Johnson *et al.*, 2012; Agegnehu *et al.*, 2015).

Additional research is still needed for clarifying the role of biochar in capturing the nutrients from urine and faeces and their release to the soil and plants. For example, pumpkin fertilized with biochar soaked in cow urine had a threefold higher yield compared to urine-only fertilized plants (Schmidt *et al.*, 2015). In addition, in a field experiment on a tropical Andosol, a fertilizer composed of co-composted pasteurized faeces, mineral additives (e.g. ash and brick particles), kitchen waste, urine and biochar gave a 16 % increase in grain corn yield compared to an unamended control. It also increased the soil available phosphorous content in a soil suspension from 0.5 to 4.4 mg kg^{-1} and the soil pH from 5.3 to 5.9, thus reducing acidification (Krause *et al.*, 2015b).

2.8 Conclusions

This chapter discussed the potential of the application of a low cost technique to treat excreta for safe and efficient recycling of nutrients and carbon matter back to the soil. It contributes to the overall approach of sustainable sanitation in which "safe disposal" transitions to safe reuse of excreta and human excreta is valued as an important fertilizer to help mitigate the current rapid global degradation of soils, while also reducing contamination of aquatic ecosystems. Lactic acid fermentation can contribute to pathogen reduction, prevent nutrient loss and control odour, thus enhancing the value of excreta as fertilizer. Lactic acid fermentation can enhance excreta hygienization, prevent nutrient losses and reduce odour, thus enhancing the fertilizing value of excreta.

The combination of lactic acid fermentation and composting (thermophilic composting or vermi-composting) and biochar addition can improve the pathogen reduction, availability of nutrients and stability of organic matter. Within this process, biochar enhances the thermophilic composting, preserves more nutrients in the substrate and increases the humus formation. The limited agricultural studies on the effects of applying lacto-fermented excreta demonstrate ample benefits on both crop production and soil quality. Among the main limitations of excreta lactic acid fermentation is the need to provide an enabling environment for LAB growth by inoculating them, e.g. as sauerkraut, and adding easily fermentable carbohydrate sources such as molasses and wheat bran. Identification of non-value carbon sources is an important consideration.

Chapter 3. Treatment of source-separated human faeces via lactic acid fermentation combined with thermophilic and vermi-composting for agricultural application

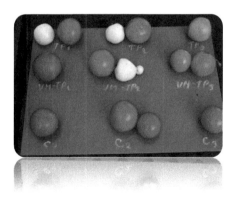

This chapter is based on:

Nadejda Andreev, Mariska Ronteltap, Boris Boincean, Piet N.L. Lens. Treatment of source-separated human faeces via lactic acid fermentation combined with thermophilic and vermi-composting for agricultural application. Compost Science & Utilization (2017, in press, 10.1080/1065657X.2016.1277809).

Abstract

Human faeces from urine diverting dry toilets can serve as valuable soil conditioners. For a successful agricultural application, an efficient pathogen reduction needs to be ensured, with no negative effects on the plants. This study assessed the efficiency of lactic acid fermentation combined with thermophilic composting on pathogen removal from human faeces and the post-treatment effects on germination and growth of radish (*Raphanus sativus*) and tomatoes (*Lycopersicum esculentum*) compared to lactic acid fermentation combined with vermi-composting and the control. The NH_4^+-N/NO_3^--N ratio of 2.99 and 3.6, respectively, indicated the obtained compost and vermi-compost was not yet mature. A complete reduction in the concentration of all investigated bacterial indicators (i.e. coliforms, *Escherichia coli*, *Enterococcus faecalis* and *Clostridium perfringens*) from 5-7 log CFU g^{-1} to below the detection limit (< 3 log CFU g^{-1}) was achieved after lactic acid fermentation combined with thermophilic composting. Lactic acid fermentation combined with vermi-composting also contributed to pathogen die-off, but coliform bacteria were reduced to only 5 log CFU g^{-1}. Compost obtained after lactic acid fermentation combined with thermophilic composting to seeds of radish had a higher germination index than vermi-compost obtained after lactic acid fermentation combined with vermi-composting (90 versus 84%). Moreover, significantly bigger average fruit weight and total biomass per tomato plant (P<0.05) were obtained after compost amendment compared to vermicast or the unamended control.

3.1 Introduction

Urine diverting dry toilets (UDDTs) offer many environmental benefits among which reduced waste volume as a result of waterless operation, simplicity of technological treatment and the possibility of recycling nutrients and organic matter to the soil (Langergraber and Muellegger, 2005; Rieck *et al.*, 2012). In UDDTs, urine and faeces are collected and contained separately, with a possibility of co-treatment with other types of organic wastes, e.g. kitchen waste or cattle manure. The separation toilet technology was successfully implemented in communities with high water tables, rocky areas or areas with no possibility to connect to the sewerage network, e.g. in regions as Central, Eastern and Northern Europe, Caucasus and Central Asia, Africa, China and Central America (Wendland *et al.*, 2011; Rieck *et al.*, 2012). During the last years, there is a growing interest towards this technology due to increasing pressures from population growth, decline of freshwater resources and soil depletion of nutrients and carbon matter. In spite of its high potential, UDDTs still need further improvements, for example on faeces sanitization and stabilization, reduction of odour as well as prevention of nutrient and carbon matter loss.

The WHO guidelines (WHO, 2006) recommend long-term storage as the simplest method of faeces sanitization. However, conditions required for pathogens or parasites die-off such as moisture reduction below 25%, pH raise above 9 or temperature increase over 55°C are seldom or never achieved in UDDTs faeces chambers in real situations (Redlinger *et al.*, 2001). Moreover, the heterogeneity of the mixed faeces and cover material can lead to a re-growth of some pathogens (Niwagaba *et al.*, 2009a). Extended storage can also cause significant loss of nitrogen and organic matter. For example, in composting toilets up to 66-80 % of nitrogen and 75% of carbon can be lost from stored excreta (Zavala *et al.*, 2002; Zavala *et al.*, 2005; Hotta and Funamizu, 2007).

Thermophilic composting and vermi-composting contribute to faecal matter sanitizing and stabilization (Eastman *et al.*, 2001; Vinneras *et al.*, 2003; Niwagaba *et al.*, 2009c; Yadav *et al.*, 2010; Juarez *et al.*, 2011; Hill and Baldwin, 2012). The main problems with thermophilic composting are the long-duration and the need to provide frequent turning for oxygen supply (Niwagaba *et al.*, 2009 b). In order to hygienize effectively the mix of source separated faeces and food waste, a thorough mixing and prolonged exposure to high temperatures is necessary during thermophilic composting (Niwagaba *et al.*, 2009c). Vermi-composting results in a homogenous end-product with high nutrient quality (Mupondi *et al.*, 2010). However, it does not always lead to a complete destruction of pathogens due to insufficient heating. This technique contributed to pathogen reduction in sewage sludge and source separated faeces (Eastman *et al.*, 2001; Buzie-Fru, 2010; Yadav *et al.*, 2010), but not always sufficiently effective for decay of resistant pathogens such as *Ascaris* (Bowman *et al.*, 2006; Hill *et al.*, 2013). Vermi-composting is also a lengthy process, e.g. 4-6 months (Alidadi and Shamansouri, 2005; Bowman *et al.*, 2006; Yadav and Garg, 2011), which leads to high losses of nitrogen and carbon matter as NH_3, N_2O and CO_2 (Frederickson and Howell, 2003; Nasiru *et al.*, 2014).

Lactic acid fermentation leads to deactivation of different types of pathogens such as faecal coliforms, *Salmonella*, *Staphylococcus* or *Clostridium* due to pH reduction as well as production of lactic acid and other compounds with suppressive effects to pathogens, e.g. bacteriocins, glucose oxidase or hydrogen peroxide (Rattanachaikunsopon and Phumkhachorn, 2010; Saranraj, 2014). A number of studies (Juris *et al.*, 1997; Papajova *et al.*, 2000; Caballero-Hernández *et al.*, 2004) showed that lactic acid fermentation may not allow a complete hygienization. For example, *Ascaris* survived after 1-6 months of lactic acid fermentation of swine faeces or silage. Another concern, besides insufficient hygienization, is the incomplete organic matter decomposition. The low pH and lactic acid formation during lactic acid fermentation inhibits the growth of microorganisms, including the growth of LAB themselves, thus promoting preservation of faecal material and inhibiting further decomposition (Murphy *et al.*, 2007). An insufficiently decomposed material applied to the soil will still have high microbial activity and may potentially deplete soil oxygen and nitrogen (Gómez-Brandón *et al.*, 2008) causing its deficiency to plant growth.

Combined lactic acid fermentation and vermi-composting has recently been considered within the terra preta sanitation approach aimed at improving faeces stabilization (Factura *et al.*, 2010; Scheinemann *et al.*, 2015). One limitation for combined lactic acid fermentation and vermi-composting is the need to pre-condition the lacto-fermented material as this is highly anaerobic and rich in organic acids, potentially negatively affecting earthworms (Buzie and Körner, 2014). In addition, studies have shown that combined lacto-fermentation and vermi-composting of human faeces was not efficient in the elimination of *Salmonella* sp. (Stoeckl *et al.*, 2013).

One possibility for better hygienization would be to combine lactic acid fermentation with thermophilic composting. The effect of integrated lactic acid fermentation and thermophilic composting on source separated faeces has thus far not been investigated. Application of lactic acid fermentation at the stage of faeces collection contributes to odour reduction and pre-processing of waste (Windberg *et al.*, 2013; Yemaneh *et al.*, 2013). Biochar is also added at the end of the lactic acid fermentation in order to reduce the nutrient loss and increase the stability of the carbon matter (Windberg *et al.*, 2013; Glaser, 2015). The enzymes and metabolites released by cellolytic and hydrolytic activities of LAB contribute to substrate solubilization and transformation, which facilitate the activity of bacterial and fungal consortia in the thermophilic composting stage (Morrison, 1988; Xavier and Lonsane, 1994; Rasool *et al.*, 1996; Yang *et al.*, 2006). The use of LAB or other effective microorganisms in combination with easily soluble carbohydrates (e.g. molasses) contributes to an increase in sanitizing temperature above 55°C and maintain it for an adequate number of days for elimination of most of the pathogens (Ong *et al.*, 2001; Ma *et al.*, 2010).

The goal of this research was, therefore, to assess the effects of combined lactic acid fermentation and thermophilic composting on pathogen reduction in faecal matter and biological effects of compost on plants compared to lactic acid fermentation and vermi-composting as well as the control. The maturity of the obtained compost and vermicompost was assessed by analysing the ammonium and nitrate ratio and the toxicological effects on radish *(Raphanus sativus)* germination and tomato (*Lycopersicum esculentum*) growth.

3.2 Materials and Methods

3.2.1. Experimental set-up

The faeces used in the experiments were collected over a three month period from a household UDDT in Chisinau (Moldova). Toilet paper was collected separately and stored together with kitchen waste in an enclosed 60 L barrel at ambient temperature (27-30 °C during August 2013 and 28-31 °C during August 2014). The experimental conditions are described in Table 3.1. After collection, the faeces were mixed with kitchen waste and cattle manure, molasses and microbial inocula. This mixture was lacto-fermented in an enclosed barrel for a period of 10 days at ambient temperature, the pH at the end of the experimental period was 5.5. After this pre-treatment, 10 % of biochar (wet weight) was added to the lacto-fermented material and further processed via either thermophilic composting or vermi-composting. Biochar was added as a bulking agent to stabilize the lacto-fermented material and increase its aeration. The experiments were repeated in two consecutive years (2013 and 2014) and during 2015 with some modifications. The changes consisted in replacing molasses by pre-fermented kitchen waste and no addition of LAB inoculum.

Chemical and microbiological analysis was done in duplicates, germination tests and growth tests were performed in triplicates. Samples for microbiological and chemical analysis were collected from 3 points and mixed together with a plastic shovel to form a representative sample. Out of this, subsamples (in duplicates) of approximately 100 g were taken, placed into plastic bags with zippers and transported to the laboratory for analysis. The composition of the mixture is presented in Table 3.1.

Table 3.1 Description of the experimental conditions and measured parameters

Time of experiment	Cover material	Composition (based on wet weight, %)	Processes	Parameters investigated
2013	Sawdust	Source separated faeces (40 %), cattle manure (30 %), fruit waste (17 %), molasses (8 %), lactic bacteria inoculum (5 %)	LAF+TC LAF+VC	T (°C), microbiological indicators Microbiological indicators Seed germination test
2014	Sawdust:pressmud 3:1 by volume	As the 2013 experiment	LAF+TC LAF+VC	T (°C), microbiological indicators, plant growth test; NH_4^+-N/ NO_3^--N
2015	Sawdust: pressmud: biochar 1:1:1 by volume	Faeces (40%), cattle manure (30%), kitchen waste (30%)	LAF+TC	Microbiological indicators, plant growth test NH_4^+-N/ NO_3^- -N, T (°C)

LAF - lactic acid fermentation and thermophilic composting, TC - thermophilic composting, VC - vermi-composting.

Table 3.2 The description of the composition of the fertilizers added

Fertilizer applied	Composition description
LAF	Source separated faeces with sawdust as cover material (40 %), 6 month stored cattle manure (30 %), fruit waste from fresh juice boutiques (17%), molasses (8%), microbial inoculum from sauerkraut juice (3 %).
LAF+TC	As above, but the addition of biochar soaked in urine for a period of two weeks
LAF+VC	Lacto-fermented mix and biochar soaked in urine were vermi-composted for 4 months with *Eisenia foetida*

LAF- lactic acid fermentation, TC- thermophilic composting, VC- vermi-composting,

3.2.2 Composting

Composting was carried out during August (2013), August-September (2014) and April-May (2015).

3.2.2.1 Thermophilic Composting

The thermophilic composting reactor was an insulated metallic box made from an old fridge (95x58x45 cm). It was positioned aslope so that the leachate could be collected during the composting period. The leachate was removed as it was accumulated and discarded. No microbiological analysis or biological test on tomato plants was performed on the leachate. Temperature was recorded once a day as an average of measurements taken from four points of the composting reactor, using a portable digital thermometer. The average outdoor temperature at the time of measurement was the following: a) 2013 - 27.8 °C (with a range of 20-33 °C), 2014 - 26.5°C (with a range of 21-30 °C) and 2015 - 22.8 °C (with a range of 18-29 °C).

3.2.2.2 Vermi-composting

Vermi-composting was carried out outdoor, in a windrow of one meter wide by one meter long and 40 cm high, using *Eisenia foetida*, obtained from the Institute of Biotechnology in Zootechnology and Veterinary Medicine (Moldova) with an inoculation density of 5,000 worms per investigated windrow and the whole experimental period lasted 130 days. Considering that the lacto-fermented material was highly anaerobic and rich in organic acids (lactic and acetic acids), with a potential deleterious effect on earthworms (Frederickson and Knight, 1988), the lacto-fermented mix was kept for one week for aeration and volatilization of toxic compounds, before being offered to the earthworms. Wet shredded newspaper was used as bedding material.

3.2.2.3 Germination and plant growth experiments

Aqueous compost extracts were prepared by mixing the compost samples with distilled water at 1:10 w/w, shaking for 1 hour and then filtered via Whatman pads (Zucconi *et al.*, 1981). For germination tests, 20 radish seeds were placed on a Whatman paper in Petri dishes and treated with aqueous extract and dechlorinated water as control. After 72 hours, germination was stopped by adding 1 ml alcohol (50%) to each of the Petri dishes. The germination index was calculated according to Tiqua *et al.* (1996).

Tomatoes were sown in plastic pots, three seeds per pot. The seedlings were transplanted after germination, one seedling per pot, into a similar soil and compost mix as they were germinating in. The total amount of nitrogen corresponded to 90 kg ha^{-1}, the quantity recommended for tomatoes growth (Andries *et al.*, 2013), the necessary quantity being calculated in accordance with the total nitrogen (N) content as g per pot, according to the methodology on pot experiments (Johnson, 2005). The compost was mixed with cernoziom clay loamy soil, the control was without any compost amendment.

The tomato plants were cultivated for a period of approximately 3 months (09.06.2014 to 13.09.2014). Watering of plants was done daily. The plant height and stem diameter were measured with a ruler, after 1.5 months from sowing and at the end of the growing period. In addition, the wet and dry weight of the root, stem, leaves and fruits as well as total biomass were weighed. The dry weight was measured by drying the material in the oven at 105° C until a constant weight was attained. Any significant difference between the means was assessed by using Dunnet multiple comparisons and one-way ANOVA with the Minitab 17 Statistical software.

3.2.3 Microbiological analysis and analytical methods

Total coliforms, *Escherichia coli*, *Enterococcus faecalis* and *Clostridium perfringens* were determined using serial dilutions of the compost and subsequent incubation on selective and differential nutrient media according to the standard protocols for pathogen and indicator microorganisms for soil and compost samples (MHRF, 2005): a) the coliform bacteria on Tergitol-7 Agar at 37°C for 24 h, b) *Escherichia coli* - on HiChrome Coliform agar at 43°C for 24 h, c) *Enteroccocus faecalis* - on Slanetz Bartley agar TTC at 37°C for 48 h and d) - *Clostridium perfringens* on iron sulphite agar at 44 C° for 16-18 h. For *Clostridium*, the samples were preliminarily heated to 75 °C for 25 minutes to eliminate the vegetative forms, after which they were sequentially heated to 70 °C and cooled in water under anaerobic conditions. The NH_4^+-N concentration was determined photocolourimetrically according to GOST 26489-85 and the NO_3^--N concentration ionmetrically according to GOST 27894.4-88 (1989).

3.3 Results

3.3.1 Temperature development during thermophilic composting

Pre-processing of the faeces and biowaste mix via lactic acid fermentation promoted a rapid increase in temperature during thermophilic composting. Figure 3.1 shows that only a very short mesophilic period was encountered during the 1ˢᵗ year (2013). The thermophilic phase, with temperature rises above 50°C started already on the second and third day. During the second year, a temperature increase above 40 °C was achieved during the 5th day and above 55 °C during the 6th day. The sanitization temperature of 56-65°C was reached at the 3rd-6th day and was maintained for a period of 9 days during 2013 and for 6 days during 2014. The sanitization temperature during 2013 and 2014 was attained without any turning or mixing, despite the relatively compacted, mostly anaerobic starting material obtained by the lactic-acid fermentation of the faeces. In contrast, during the year 2015, the sanitizing temperature was not attained: the overall active stage of composting occurred mainly at a mesophilic temperature (below 40 °C) (Figure 3.1). The temperature increase above 50 °C took place only after the addition of a small proportion of molasses (1%), but it was maintained for three days only, thus showing a quick depletion of this available carbon source. It might be also that the outdoor temperature had an influence on the compost temperature.

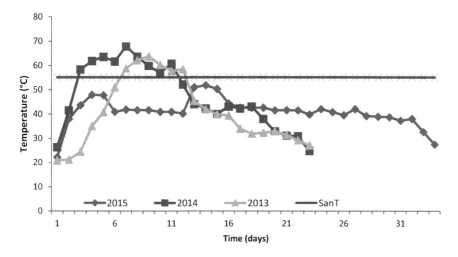

Figure 3.1 Temperature variations during thermophilic composting after pre-processing of faecal material and biowaste via lactic acid fermentation. SanT - Sanitization temperature

The decrease in temperature below 50 °C and transition to the curing phase took place after 12 days and lasted for 11 days during 2013. In 2014, the transition to the curing phase occurred after 14 days of thermophilic composting and lasted for 11 days as well. At the end of this stage, the temperature became similar to the ambient temperature, i.e. 25-27°C. The overall composting period during 2013 and 2014 lasted for 23 days, but when also considering 10 days from the lactic acid fermentation, it lasted for 33 days. During 2015, with no pre-treatment via lactic acid fermentation, the overall composting period lasted for 33 days. The values of the NH_4^+-N/NO_3^--N ratio was 2.99 for the thermophilic compost and 3.6 for the vermi-compost, indicating that the compost was not yet mature (Bernal *et al.*, 2009) (Table 3.3). Additional research is required to assess the maturation and stabilization time required.

Table 3.3 Nitrate and ammonium content in the compost (Mean ± SD, n = 2)

Sample type	NO_3^--N, mg kg^{-1}	NH_4^+-N, mg kg^{-1}	NH_4^+-N/NO_3^--N	pH
LAF+TC	334 ± 4.2	1000 ± 14.8	2.99	7.78. ± 0.2
LAF+VC	277.6 ± 17.7	1000 ± 14.1	3.6	7.97 ± 0.1

Even though no measurements were made on the decomposition of lignocellulotic materials, visual observations showed that at the end of the thermophilic composting period, there were almost no signs of toilet paper in the compost anymore; the compost was covered by white mould.

3.3.2 *Hygienization by lactic acid fermentation combined with thermophilic composting or vermi-composting*

Table 3.4 shows that the lactic acid fermentation stage contributed to a reduction of all pathogen indicator organisms by 2-4 log_{10} CFU g^{-1}. According to EPA pathogen reduction requirements for class A biosolids, the level of coliform bacteria should be reduced to < 3 log CFU g^{-1} for a safe agricultural application (Pepper *et al.*, 2006). Lactic acid fermentation alone did not reduce the pathogen indicators to this level. The concentrations of microbiological indicators during this process were reduced as follows: coliforms from 7 to 3.9-4.5 log CFU g^{-1}, *E.coli* from 4.6-6.7 to 3.2-4.5 log CFU g^{-1}, *E. faecalis* from 6.7-7.2 to 5-5.5 log CFU g^{-1} and *Clostridium perfringens* from 5 to 2.5-4.5 log CFU g^{-1} (Table 3.4).

Instead, lacto-fermentation combined with thermophilic composting decreased the number of all four pathogen indicator organisms from 5-7 log CFU g^{-1} to the recommended safe level below 3 log CFU g^{-1}. Lactic acid fermentation combined with vermi-composting also led to pathogen reduction, however, not to the safe level of 3 log CFUg^{-1} for coliform bacteria both during 2013 and 2014.

Table 3.4 - Density of pathogen indicator organisms in faeces: biowaste mix before treatment, after lactic acid fermentation and combined lactic acid fermentation/thermophilic composting or vermi-composting (average values of the samples analysed during 2013-2014).

Pathogen indicators	Bacterial density log$_{10}$ CFUg^{-1}			
	Raw material	LAF[1] (10 days)	LAF+TC[2]	LAF+VC[3]
2013				
coliforms	7.09±0.5	3.9±2.6	2.7±0.6	5.2±1.0
E. coli	6.7±1.2	3.2±1.6	2.3±0.6	2.0±0.0
E. faecalis	7.2±0.4	5.0±1.7	2.0±0.0	3.7±0.0
C. perfringens	5.0±0.0	2.5±0.7	2.0±0.0	2.0±0.2
2014				
coliforms	5.3±0.5	4.5±2.1	3.0±0.0	4.9±0.5
E. coli	4.6±0.0	4.5±2.1	3.0±0.0	3.0±0.0
E. faecalis	6.7±0.0	5.5±0.2	2.0±0.0	2.0±0.0
C. perfringens	1.0±0.0	4.5±0.7	1.5±0.7	2.3±1.0

[1]LAF - lactic acid fermentation; [2]LAF+TC - combined lactic acid fermentation and thermophilic composting; [3]LAF+ VC - combined lactic acid fermentation and vermi-composting.

3.3.3 Effects on plant germination and growth

Both applied fertilizing mixes had beneficial effects on radish germination (GI > 80%). However, radish seeds treated with aqueous extract of compost obtained after lactic acid fermentation combined with thermophilic composting showed a higher GI than that after lactic acid fermentation combined with vermi-composting (90 % versus 84 %). The tomatoes grown in soil supplemented with both compost and vermicast were more vigorous than in the control (Figure 3.2 A and B).

Figure 3.2 Tomato plants treated with compost obtained after lacto-fermentation with thermophilic composting (front row in photo A) and lacto-fermentation with vermi-composting (front row photo B) in comparison to the control (back row photos A and B). Three most representative pots of each treatment are presented in this picture.

There were 18% and 7 % taller tomato plants in pots fertilized by compost obtained after lacto-fermention combined with thermophilic composting and lacto-fermentation combined with vermi-compost respectively, compared to the control. After 1.5 months from sowing, the plant height of tomatoes fertilized by compost and vermicast was significantly higher than the control (Table 3.4).

Even though the root, leaf and stem weight (dry and wet weight) was not significantly different than the unfertilized control, the fruits of tomato fertilized by compost obtained after lactic acid fermentation combined with thermophilic composting were significantly bigger than those of the control. In addition, this type of compost produced the highest amount of tomato biomass (wet and dry weight) compared to vermicast or the control (Table 3.5).

Table 3.5 Effect of compost versus vermicast on growth and weight of tomatoes

Sampling period	Parameter	Control	LAF/VC	LAF/TC
1.5 months after sowing	PH[1] (cm)	4.6±0.8 A	6.63±1.0 B	8.2±1.0AB
End of experiment 1.5 month after sowing.	PH (cm) SD2 (cm)	39.6±0.5 A 0.84±0.14 A	37.6±3.5 A 0.90±0.0 A	41.8±1.04 A 1.07±0.15 A
End of experiment	SD (cm)	1.45±0.02A	1.72±0.04A	1.96±0.02 A
End of experiment	Rww3 (g)	1.74±0.9 A	1.4±0.15 A	2.29±0.2 A
End of experiment	Rdw4 (g)	0.67±0.1 A	1.14±0.6 A	1.24±0.5 A
End of experiment	Lww5 (g)	8.03±1.6 A	7.25±2.1 A	8.04±1.8 A
End of experiment	Ldw6 (g)	2.24±0.6 A	2.4±0.1 A	2.73±0.7 A
End of experiment	Sww7 (g)	11.73±0.4 A	11.8±1.7 A	15.2±2.1 A
End of experiment	Sdw8 (g)	2.37±0.5 A	2.5±0.4 A	3.2±0.4 A
At harvest	Fww9/plant (g)	31.±3.3 A	41.4 ±1.9AB	55.1±12.5 B
At harvest	Fdw10 (g)	2.37±0.5 A	2.5±0.49A	3.17±0.4 A
End of experiment	TBww11 (g)	52.5±2.5 B	62.6±4.9 B	78.91±11.3 A
End of experiment	TBdw12 (g)	7.97±0.84 B	11.45±1.23 B	8.9±0.9 A

[1]-plant height; [2]- stem diameter; [3] - root wet weight; [4]- root dry weight; [5]- leaf wet weight; [6]- leaf dry weight; [7]- stem wet weight; [8] - stem dry weight; [9] - fruit wet weight; [10] - fruit dry weight; [11]- total biomass wet weight (per plant), [12] - total biomass dry weight (per plant). The means were compared for significance using Fisher Pairwise comparison (p<0.05). Means that do not share a letter are significantly different.

3.4 Discussion

3.4.1 Combined lactic acid fermentation and thermophilic composting

This study showed that lactic acid fermentation combined with thermophilic composting during a two year run experiment might contribute to an overall shorter composting period (33 days) compared to the usual aerobic composting (45-180 days) (Wang *et al.*, 2007; Fornes *et al.*, 2012). Thermophilic temperatures in the current study were achieved during the second-fifth day without any manual turning with passive aeration only, such type of compost being described as of the "micro-aerobic type" (Wang *et al.*, 2007; Green and Popa, 2011; Alattar *et al.*, 2012), however to prove this determination of the oxygen concentration in the compost pile is required.

The overall composting period of 33 days (together with the lactic acid fermentation period) is similar to other studies (Wang *et al.*, 2007). It would be interesting to investigate how thermophilic and vermi-composting processes alone differ in their rate of decomposition and stabilization of organic matter and nutrient content compared to the integrated processes of LAF and thermophilic or vermi-composting. The fast temperature increase without any turning was probable related to the addition of molasses, a source of labile carbon, which boosted the number of thermophilic bacteria (Hayes and Randle, 1968). In addition, the use of lactic acid fermentation as pre-processing stage most probable contributed to the partial decomposition and solubilization of excreta, thus facilitating more efficient substrate utilization during thermophilic composting. This might have been caused by the abundance of enzymes released by the LAB (Morrison, 1988; Rasool *et al.*, 1996; Ong *et al.*, 2001; Yang *et al.*, 2006). It is important to understand the contribution of lactic acid bacteria and molasses (or other sources of water soluble carbohydrates) to the transformation of faecal substrate prior to thermophilic composting.

Based on the assessment of the NH_4^+-N/NO_3^--N ratio, the obtained compost was not yet mature, higher than the indicator value for compost maturity of > 0.16 (Bernal *et al.*, 1998). For a more complete picture of compost maturity, other parameters should be investigated, e.g. C/N ratio, respiration rate (Barrena *et al.*, 2014), or the level of elemental and functional composition of organic matter and the humification level (Bernal *et al.*, 2008) using electrophoresis, chromatography-mass spectrometry or UV-spectroscopy (Domeizel *et al.*, 2004; Bernal *et al.*, 2009).

The overall aerobic composting takes longer than in micro-aerobic composting, e.g. 45-50 days versus 35 days. With the exclusion of turning and mixing, it is also possible to reduce carbon and nitrogen losses as CO_2 and NH_3 (Liang *et al.*, 2006). As the process was micro-aerobic, thermophilic bacteria probably contributed to the partial decomposition of the toilet paper during the thermophilic stage. In addition, enzymatic systems of lactic acid fermentation bacteria might have added to the decomposition of celluloses and hemicelluloses into simpler compounds during the lactic acid fermentation stage (Morrison, 1988; Rasool *et al.*, 1996).

3.4.2 Faeces sanitizing by combined lactic acid fermentation with composting

This study showed that integrating the processes of lactic acid fermentation with thermophilic composting was more efficient in the reduction of coliform bacteria and *E. faecalis* than combining lactic acid fermentation with vermi-composting (Table 3.3). As the pH was 5.5 only at the end of lactic acid fermentation, that was probably not the single factor contributing to pathogen reduction. Also antimicrobial substances produced by LAB during the fermentation might have contributed to pathogen reduction. For example, *Lactobacillus plantarum*, which was applied in this study, produces bacteriocins, that are thermoresistant and inhibit a wide range of pathogens. The bacteriocins, might have been broken down by proteases produced in higher amounts in vermi-composting than in thermophilic composting (Abo-Amer, 2007; Devi *et al.*, 2009). The main mechanisms of pathogen reduction during thermophilic composting are: competition between indigenous microorganisms and pathogens, the antibiotic activity of excretions of fungi or actinomycetes, natural die-off and thermal destruction (Burge *et al.*, 1978; Haug, 1993). The most important factor in pathogen destruction is, however, the increased temperature (Vinnerås *et al.*, 2003). According to the sanitization standards for biosolids (Noble *et al.*, 2004; Schönning and Stenström, 2004), a minimum temperature of 55°C for a period of 3-14 days is required for achieving a safe product for agricultural application, depending on the heterogeneity of the compost material. During this study, hygienization temperatures above 55 °C were maintained for 6-9 days, which was likely sufficient to reduce the bacterial contamination to a safe level of below the detection limit of < 3 log CFU g^{-1}.

The mechanism of pathogen reduction in vermi-composting is caused by the direct destruction of the pathogenic bacteria after passage via earthworm guts and under the effect of coelomic fluids or glycoproteins excreted by the earthworms (Edwards *et al.*, 1984; Popović *et al.*, 2005). The insufficient hygienization during vermi-composting was also mentioned by other studies (Alidadi *et al.*, 2007), which showed that vermi-composting reduced pathogens to a level which satisfies EPA requirements for class B biosolids, but not for class A. In this study during lactic acid fermentation combined with vermi-composting, at a density of 5000 earthworms/m^2 there was insufficient reduction of coliforms and in some cases of *Enterococcus faecalis*. Other studies reported a complete pathogen removal at a density 6-9 times higher than that used in this study (Eastman *et al.*, 2001). Such a high density, if applied at large scale, might not be economically feasible (Mupondi *et al.*, 2010).

It would be interesting to study the effect of lactic acid fermentation combined with thermophilic composting on resistant pathogens such as *Ascaris*. Lactic acid fermentation can reduce the viability of *Ascaris suum* in swine faeces; however, it did not destroy them completely even after 56 days of lactic acid fermentation (Caballero-Hernandez *et al.*, 2004). Scheinemann *et al.* (2015), has achieved, however, a succesful *Ascaris* elimination after 56 days of lactic acid fermentation, but only at temperatures above 36 °C.

3.4.3 The value of faecal compost obtained via combined lactic acid fermentation with thermophilic composting as soil amendment

The germination index (>80%) of both the thermophilic compost and vermi-composts showed good potential for improving plant growth when used as soil amendments. Studies have reported that vermi-composts can be efficient soil amenders, with a higher quality than normal compost (Chaoui *et al.*, 2003; Tognetti *et al.*, 2005). For example, while a total replacement of the peat growth media with vermi-compost was possible, an increase in compost dose above 50 % caused plant mortality (Lazcano *et al.*, 2009). One of the explanations could be that the nitrogen is released in the form of ammonium during composting, which may be toxic to plants, while in the vermi-compost it was in the nitrate form, which is plant available (Atiyeh *et al.*, 2000).

This study showed that lactic acid fermentation with thermophilic composting had an overall higher effect on germination of radish *Raphanus sativus* and growth of tomatoes *Lycopersicum esculentum* than lactic acid fermentation with vermi-composting (Table 3.5). Combined lactic acid fermentation with thermophilic composting can be a faster turnover rate of human faeces than lactic acid fermentation with vermi-composting (33 versus 130 days), therefore with potential to recycle more nutrients and organic matter. Most probably these integrated processes produced a more stabilized end-product, with a higher nutrient content than the lactic acid fermentation and vermi-composting. However, it did not have any negative effects on germination, the NH_4^+-N/NO_3^--N ratio indicated that the compost was not yet mature. Table 3 shows that the nitrate content in the compost was higher than in the vermi-compost. This may be related to the fact the biochar added during the thermophilic stage contributed to ammonia and nitrate retention via sorption (Knowles *et al.*, 2011; Jindo *et al.*, 2012; Hijikata *et al.*, 2015), thus reducing the potential toxic effects of ammonia to the plants.

3.5 Conclusions

Lactic acid fermentation combined with thermophilic composting can be a cost-effective method for the processing of source separated faeces from UDDTs for agricultural reuse that could be performed without turning but with only passive aeration. This integrated process was more efficient in the reduction of pathogen indicator bacteria and had a better stimulatory effect on the germination of radish and growth of tomatoes and occurs at a shorter turnover rate (33 days) than lactic acid fermentation combined with vermi-composting (130 days).

Chapter 4 The effect of lacto-fermented faeces, biowaste and addition of biochar soaked in urine on soil quality, growth, yield and yield components of *Zea mays* L.

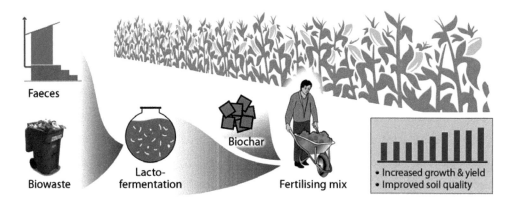

This chapter is based on:

Nadejda Andreev, Mariska Ronteltap, Piet N.L. Lens, Boris Boincean, Lidia Bulat, Elena Zubcov (2016) Lacto-fermented mix of faeces and bio-waste supplemented by biochar improves the growth and yield of corn (*Zea mays* L.) Agriculture, Ecosystems and Environment, 232, 263-272

Abstract

Resource oriented sanitation has emerged as a need to improve the efficacy of excreta treatment schemes, to reduce the environmental pollution from their disposal and to improve soil fertility. In urine diverting dry toilets, storage alone is inefficient for faeces treatment due to poor hygienization, incomplete decomposition as well as high losses of organic matter and nutrients. The purpose of this study was to assess the effect of stored faeces and bio-waste treated via lactic acid fermentation supplemented with biochar soaked in urine on the growth, yield and yield components of corn in a two year field experiment. Also the soil quality was assessed. Lactic acid fermentation of faeces and bio-waste was performed under limited oxygen conditions in earth pits covered with plastic foil by the addition of microbial inoculum and molasses. The effect of fertilizer was compared to an unfertilized control; a lacto-fermented mix (faeces, bio-waste, molasses and microbial inoculum) without biochar; stored faeces; cattle manure; urine and mineral fertilizer. In addition, during the second year, a vermi-composted lacto-fermented mix supplemented with biochar was applied. Differences among the treatments concerning corn growth parameters, yield, yield components and soil quality were evaluated using the Dunnet test of multiple comparisons. The lacto-fermented mix supplemented by biochar significantly improved plant height ($p < 0.05$ and confidence interval CI with two negative values) compared to all fertilizers during the first production year and compared to the control, stored faeces and vermicompost during the second year. This fertilizer also achieved a significantly higher corn yield compared to all other fertilizers during the first and second production year, except for the lacto-fermented mix without biochar and the mineral fertilizer, which showed no significant yield and yield component difference ($p > 0.05$, CI with both positive and negative values). With regard to the soil quality, bulk density was reduced during both years, while the soil potassium content increased during the first production year. The nitrate concentration significantly improved in comparison to the control, stored urine and stored faeces amended plots.

4.1 Introduction

Adequately treated human faeces from urine diverting dry toilets (UDDTs) can serve
as an important soil conditioner. Studies have demonstrated a number of benefits from stored
human faeces such as improving the soil structure, water holding capacity, soil microbial
activity, soil buffering capacity and nutrient use efficiency (Guzha *et al.*, 2005). In UDDTs,
faeces have to be maintained dry and stored for one to two years, without addition of new
fresh faeces, in order to eliminate or substantially reduce the number of pathogenic organisms
(WHO, 2006). The number of infective vectors in the stored faeces decreases as a result of
natural die-off or by the influence of environmental factors such as low moisture content and
pH increase (Redlinger *et al.*, 2001; Kaiser, 2006). However, reduced moisture content
(below 40 %) leads to an incomplete decomposition, whereas a high pH contributes to intense
nitrogen volatilization (Nordin, 2010). Storage can result in the loss of about 66-94% of the
nitrogen and a significant proportion of the carbon (Zavala *et al.*, 2002; Zavala *et al.*, 2005).
In addition, storage alone may not contribute to a complete sanitization of the faeces, as there
is a potential re-growth of some pathogens (Gibbs *et al.*, 1997). This method may thus
provide a product with an incomplete decomposition, still unsafe to be applied in agriculture
and with a relatively low nitrogen and carbon content (Nordin, 2010).

Application of safely treated human excreta to agriculture, rather than land disposal,
supports the reduction of environmental pollution and allows for recycling of valuable
components such as nitrogen, phosphorus and organic matter to the soil (Murray and Buckley,
2010b; Lu *et al.*, 2012). Fertilizing crops, by source separated human faeces, alone or in
combination with urine contributed positively to soil quality, growth and yield of different
crops, similar to or even better than manure or mineral fertilizers (Guzha *et al.*, 2005b;
Mnkeni and Austin, 2009; Triastuti *et al.*, 2009). Adding combined human faeces and urine
resulted in a higher water use efficiency by corn than urine or mineral fertilizer only (Guzha
et al., 2005).

The application of mineral fertilizer or urine alone might be less efficient due to
nitrogen leaching. For example, a lysimeter study with ^{15}N-labeled cow urine has shown that

in a temperate climate, approximately 60 % of the nitrogen was leached within one year following urine application (Wachendorf *et al.*, 2005). Supplementation with urine and faeces does not only increase the soil nutrient content but also the soil organic matter content, which in turn raises the water holding capacity (Hatfield *et al.*, 2014). For example, field experiments showed that for every 1 % increase in soil organic matter, there is a 3 time increase in soil water holding capacity (Lawton, 2014) and an accumulation of 24 kg ha^{-1} of mineral nitrogen (Andries, 2007).

Studies have shown that human excreta in combination with biochar can improve soil quality by increasing nutrient availability and reducing soil acidification (Krause *et al.*, 2015a; Krause *et al.*, 2015b). Such studies have emerged after the discovery of highly fertile soils of anthropogenic origin such as terra preta or Amazonian dark earths (Woods, 2003; Glaser and Birk, 2012) and their analogues, e.g. in Northern Europe (Wiedner *et al.*, 2015a), Australia (Downie *et al.*, 2011) and West Africa (Frausin *et al.*, 2014).

An important prerequisite for agricultural application of human faeces is their hygienization. Among different treatment techniques, lactic acid fermentation combined with vermi-composting has been considered during the last years in the context of the so-called "terra preta sanitation" approach (Windberg *et al.*, 2013). At different stages, biochar can be added to reduce excessive nutrient loss and stabilize organic matter (Schuetze and Santiago-Fandiño, 2014). To restore soil fertility, soil management practices which maximize recycling of nutrients and build up of soil organic matter need to be adopted (Roy *et al.*, 2002; Andries, 2007). Lactic acid fermentation, also called ensilage, is a widely used technique for the preservation of food and animal feed (Kamra *et al.*, 1984; Andersson *et al.*, 1988; Schroeder, 2004). Lacto-fermenting human faeces together with a microbial inoculum and other types of organic waste such as cattle manure, molasses or vegetable residues can return a larger amount of carbon and nitrogen to the soil than stored faeces or cattle manure alone because their losses are reduced. For example, 58 % of organic matter is lost in the common compost process, compared to only 2.5% in fermented material (Hitman *et al.*, 2013).

Lactic acid fermentation of faeces eliminates the foul odour and suppresses the growth of putrefactive bacteria (Hata, 1982; Freitag and Meihoefer, 2000; Wang *et al.*, 2001). This type of treatment proved to be an efficient method in reducing the number of pathogens in manure and sewage sludge (Scheinemann *et al.*, 2015). Reduction of pH during lactic acid fermentation reduces nitrogen losses as NH_3 (Berg *et al.*, 2006) because the activity of lactic acid bacteria inhibits urea hydrolysis, thus preventing the volatilization of ammonia (Larson and Kallio, 1954; Schneider and Kaltwasser, 1984). Direct application of lacto-fermented faeces may raise some problems for the soil. This is because the end-product of lactic acid fermentation is anaerobic and rich in available carbon, thus its addition may intensify the activity of soil decomposers and mineralization of soil organic matter (Hamer and Marschner, 2005; Green and Popa, 2011). As a consequence also the oxygen concentration in the root zone may decrease (Jiménez and Garcia, 1989).

Mixing of lacto-fermented organic waste with biochar may contribute to synergistic positive effects on nutrient level, plant available water retention and an increase in stable carbon content (Dias *et al.*, 2010; Liu *et al.*, 2012). The labile carbon diffuses into the biochar pores and is adsorbed onto the biochar surface, thus preventing abiotic and biotic organic matter decomposition (Zimmerman *et al.*, 2011). The organic matter added with the lacto-fermented substrate as well as the biochar increases the cation exchange capacity of the soil and, therefore, the adsorption and retention of NO_3^-, NH_4^+-N and NH_3 (Jindo *et al.*, 2012). Studies have shown that biochar supplemented to compost adsorbed nitrate and released it to the plants, in this way preventing nitrate leaching and thus acting as a slow release fertilizer (Kammann *et al.*, 2015).

Biochar can be enriched with nutrients from human urine (Andreev *et al.*, 2016). The small organic particles present in urine form a coating on the biochar onto which anions such as nitrates as well as cations like ammonium can bind, which favourably influence the crop yield (Kammann *et al.*, 2015; Schmidt *et al.*, 2015). For example, the addition of a mix of cow urine and biochar significantly increased the yield of pumpkin compared to the urine or biochar only treatments (Schmidt *et al.*, 2015).

Rational nutrient and organic carbon fertilizing management schemes from sanitation waste are particularly important for subsistence farmers whose agricultural benefits are exacerbated by the increase in intensity and frequency of droughts. Limited financial resources and lack of knowledge on the soil fertilization practices force farmers to use limited or no mineral and organic fertilizers (Andries *et al.*, 2014). Crop yield mainly relies on natural fertility, thus leading to loss of humus and further impoverishing the soil (Krupenikov *et al.*, 2011).

In Moldova, the decrease in precipitation and frequent droughts occurring during the last decades lead to severe declines in crop yields (Potop, 2011; Vronskih, 2014). Land privatization has led to the formation of many small plots distributed to rural farmers that are only enough for subsistence farming. While crops extract up to 150-180 kg ha^{-1} NPK, the supplementation from fertilizers is almost 10 times less, contributing to a negative nutrient balance (Andries *et al.*, 2014). Hence, a permanently available and affordable nitrogen fertilizer, such as co-treated human faeces and bio-waste, has a large potential in the agricultural future of countries like Moldova. At the moment, several urine diverting dry toilets are available all over the country. The faeces quantity is relatively small and it takes several years until the faeces compartment is filled. Faeces from these urine diverting dry toilets could be lacto-fermented together with other types of organic waste such as animal manure or waste from the food and beverage industry.

The purpose of this study was to determine whether a lacto-fermented mix obtained by combined treatment of faeces from UDDT and bio-waste, such as manure, fruit residues and waste from sugar industry (molasses) by lactic acid fermentation with the addition of biochar soaked in urine has a beneficial effect on the productivity of corn (*Zea mays* L.) and the soil quality. The influence of the applied fertilizer on the growth rate, yield and yield components of corn, nitrate, phosphorus, potassium and humus content, soil available water as well as soil bulk density, were examined during two production years. The lacto-fermented mix supplemented by biochar was compared to an unfertilized control, a lacto-fermented mix without biochar, mineral fertilizer, stored faeces, urine and cattle manure. An additional treatment step via vermi-composting after lactic acid fermentation was performed as well, but this material was only applied on the test plots during the second production year.

4.2 Methods

4.2.1 Site description

The research was carried out during two growth seasons (2012-2013 and 2013-2014) on a private farm located in the Bolduresti Village (Nisporeni District, Figure 4.1) at a latitude of 47° 8′6.72″ N and longitude of 28°2′5.85″E, 90 km North-West of Chisinau (Moldova). The annual average minimum air temperature for Nisporeni during the 2012-2013 production period was 5.52 °C and during the 2013-2014 production period it was 5.92 °C. The average annual maximum temperature was 15.75 and 16.38 °C for the first and second production period, respectively (Table 4.1). The total rainfall for the first production period was higher than the second production period (580 mm versus 547 mm, Table 4.1). The fall and spring were more favourable for the second production year, with higher precipitation during these

Figure 4.1 Location of the experimental plot in the village Bolduresti (Nisporeni district, Moldova)

seasonal periods (Figure 4.1).

Table 4.1 Average temperature and precipitation data in the Nisporeni district during the production periods

	2012				2013		
Month	Aver. max t°C	Aver. min t°C	Precip. mm	Month	Aver. maxt° C	Aver. min t°C	Precip. mm
IX	16.74	12.22	43.9	IX	20.50	9.50	93.7
X	19.00	6.20	33.8	X	16.83	6.29	1.9
XI	14.17	3.06	16.9	XI	12.80	4.60	23.6
XII	-0.58	-8.92	40.3	XII	4.88	-3.04	5.2
	2013				2014		
I	0.32	-6.13	46.3	I	1.77	-4.32	44.8
II	4.00	-1.79	21.7	II	2.48	-4.10	7.2
III	7.58	-1.74	32.9	III	14.26	2.81	15.6
IV	18.97	6.26	35.5	IV	17.20	5.66	76.2
V	25.48	11.96	84.8	V	23.03	10.22	138.4
VI	27.13	15.30	172.7	VI	25.33	13.20	29.4
VII	27.62	15.32	102.6	VII	28.00	15.81	96.4
VIII	28.55	14.52	49.9	VIII	29.48	15.06	14.7
Prod period	15.75	5.52	681.3	Prod period	16.38	5.97	547.1

Source: Anonymous, 2015.

The means of precipitation for the investigated production years were compared to the average multiannual precipitation, representing data from long term monitoring (1981-2010) collected across the country (Nedealcov *et al.*, 2013) (Table 4.2). The annual precipitation for both investigated years was higher than the average long term (30 years, 1981-2010) precipitation value for the Nisporeni district, 681 and 547 mm versus 538.9 mm, respectively (Nedealkov *et al.*, 2013) (Table 4.2).

The investigated soil was a clay-loamy cernoziom, slightly eroded, with a low humus content of approximately 2.83 % and a bonity of 66 (Baluhatîi *et al.*, 2002). Soil bonity represents a comparative qualitative assessment, based on a scale from 0 to 100 points of soil price or productive capacity and is calculated for each individual soil and compared to a reference soil (Ursu *et al.*, 2009). According to its humus content, the soil belongs to the IIIrd bonity class, which is an average value characteristic for the current soil status of Moldova (Leah and Andries, 2012). This type of soil bonity allows for an annual crop productivity of winter wheat of 2.5 t ha[-1] (National Bureau of Statistics of the Republic of Moldova, 2014). The humus value is, however, quite low compared to the average humus content in the topsoil for cernozioms in Moldova, which ranges from 3.5-4 to 5-6 % (Andries, 2006).

Table 4.2 Classification of the investigated production years (2013 and 2014) according to the amount of precipitation (mm) during spring and summer

N/o	Period	Years	Precipitation (mm) for the investigated production year	Precipitation (mm) average value during 1981-2010
1	Spring	I[st] production year	51**	44
2	Spring	IInd production year	77 **	
3	Summer	I[st] production year	108 **	67.5
4	Summer	IInd production year	47 *	

[1] - The average precipitation value is based on data for the spring and summer periods of the investigated district (Nedealkov *et al.*, 2013). * - significantly lower, **- significantly higher.

The pH of the soil varies between neutral to slightly alkaline. The investigated area was previously used mainly for corn production for several years. According to the land owner, no fertilizers were applied to the area since the land privatization (1991-1992) and the yield was mainly based on the natural fertility of cernoziom soils.

4.2.2 Fertilizers applied in this study

This study is focused mainly on a lacto-fermented stored faeces from UDDT and bio-waste combined with urine charged biochar. The combined fertilizer was compared to a lacto-fermented mix without biochar, an unfertilized control, stored faeces, cattle manure and urine as well as an NPK mineral fertilizer. In addition, during the second production year, the lacto-fermented mix supplemented by biochar was also vermi-composted, using *Eisenia foetida* earthworms. Since the entire vermi-composting process required four months for completion, a sufficient quantity for field application was only obtained and applied during the second year.

The lacto-fermented mix consisted of stored faeces, cattle manure, fruit residues, sugar beet molasses and microbial inoculum from sauerkraut brine, which were added at a proportion of 40:30:17:8:5 percent by wet weight. The one-year stored faeces were obtained from UDDTs of three schools in the Nisporeni district (Moldova). The faeces were homogenized and separated from any foreign material (for example toilet paper and hygienic pads). The cattle manure that was previously stored at ambient temperature for a period of 6 months was obtained from a farm within the village of the experimental area. The fruit residues were purchased from a fresh juice boutique in Chisinau and chopped in a grinder prior to use. Molasses was obtained from a sugar factory in Faleshti (Moldova). The sauerkraut brine was activated by mixing with sugar beet molasses and water at a ratio of 5:5:90 percent by volume and fermented for five days, which allowed microbial growth (Jusoh *et al.*, 2013). After a thorough mixing, the waste components were placed in an earth pit and fermented for a period of eight weeks. The material was tightly compacted to ensure micro-aerobic conditions, and plastic foil was used to line and cover the pit (Figure 4.2). Lactic acid fermentation in earth pits is quite a laborious process for practical applications at large scale, windrow compaction by tractor might be preferred for the latter.

Biochar residues (dust and small pieces) from a local enterprise producing wood biochar are presented in Table 4.3. The biochar proximate analysis was carried out according to reported methodologies (Ronsse *et al.*, 2013).

Figure 4.2 The earth pit where the waste material was fermented. In order to ensure microaerobic conditions, the following actions were undertaken: A) the earth pit was lined with plastic foil and the material was tightly compacted by feet and B) the pit was covered with plastic foil and soil.

Biochar samples (1 g) were crushed, placed in porcelain crucibles and oven dried at 105°C for 2 hours to determine the moisture content. The volatile matter was determined as the weight loss of biochar by heating the samples at 950 °C for 11 min in covered crucibles. The ash content was determined by heating for a minimum of 2 hours (uncovered crucibles) at 750°C. The stable carbon fraction (fixed carbon) was calculated by subtracting its moisture content, ash content and volatile matter content. The surface area and pore volume was measured by nitrogen gas absorption using a quanta chrome instrument Autosorb 1 MP (Greg and Sing, 1984).

The specific surface area of the biochar investigated (Table 4.3) was lower than that of other reported wood biochars (Ronsse *et al.*, 2013). The investigated biochar had a moisture, ash and fixed carbon content within the limits for previously reported wood biochar (Brewer *et al.*, 2011). Also the volatile matter was within the limit of variation of 20-60 % (33%), characteristic for wood biochar (Enders *et al.*, 2012).

Table 4.3 Characteristics of the quality of wood biochar used as fertilizer supplement

Moisture (%)	Volatile matter (%)	Ash content (%)	Fixed C (%)	Specific surface area (cm²/g)	Pore volume (cm^3/g)
2.35 ± 1.9	32.65 ± 1	7 ± 2.82	60.35 ± 13.24	73.79 ± 8.57	0.063 ± 0.036

Prior to the application, the charcoal was charged with nutrients from human urine by storing a mixture of the urine and charcoal (3:1) in closed plastic barrels of 35 L for a period of two weeks. Analysis of the urine solution before and after soaking showed a retention of 29 % of the ammonia and 70% of the phosphate onto the biochar (Table 4.4).

Table 4.4 Ammonia and phosphate concentration in urine before and after biochar soaking (mean ± SD, n = 3)

Urine solution	$N-NH_4^+$ (mg/l)	$PO_4^{3-}-P$ (mg/l)
Before biochar charging	9600 ± 20	520 ± 10
After biochar charging	6900 ± 10.5	150 ± 6

As mineral fertilizer, a complex nitrogen and phosphorus (amofos) fertilizer was used with a N and P content of 12% and 52 %, respectively, supplemented by ammonium nitrate (NH_4NO_3) and potassium chloride (KCl with K content of 40%). The mineral fertilizer was offerred by the Selectia Institute of Crop Research (Moldova).

4.2.3 Sample collection and frequency

The soil samples were collected one time per growth season, four weeks after planting, using a spade and a scoop. A composite sample from each plot was obtained by collecting random subsamples at depths of 0-20 and 20-40 cm. In order to ensure representativeness of the sampling depth, a ruler was used and the same collection method was applied along the whole experimental area. In each plot, three subsamples from both soil depths were collected from the middle, these being mixed in a common sample of approximately 1 kg, out of which a small subsample of 300 g was drawn for analysis.

4.2.4 Physico-chemical analysis

Bulk density was assessed by core sampling (taken in 3 points of the sampling plot) and calculated as the ratio of oven-dried mass to its volume (Stadnik, 2010). Before the chemical analysis, the soil was air-dried, crushed and sieved.

All the elemental analysis (total N, P, K and C) was done from the same soil sample. Samples for nitrate analysis were collected separately and analysed during the following day after collection. The soil was collected four weeks after planting, at a depth of 0-20 and 20-40 cm, as described in Andreev *et al.* (2016). For measuring nitrate concentrations, an ion meter I-160 M (Belarus) and an electrode ELIS 121-NO_3^- were used. The calibration of the electrode was done by using known N-NO_3^- standard solutions, made by serial dilution of a 1000 ppm solution.

The soil available water storage capacity is the difference between water at field capacity and the permanent wilting point (Eq. 1):

$$Wa \ (mm) = (Wfc - Cw) * H * d * 0.1 \quad \text{(Eq. 1)}$$

Where W_a - soil available water storage capacity (mm), W_{fc} - soil gravimetric moisture (soil water at field capacity, % C_w - permanent wilting point, % H - soil bulk density (g/cm^3), d - soil depth (cm) (Verigo and Razumova, 1973). Water at field capacity is the water remaining in the soil one or two days after rainfall. A permanent wilting point is the moisture content of the soil at which plants wilt and fail to recover when they are supplied again with sufficient moisture (Verigo and Razumova, 1973). In this study, the calculated value for clay soils (Verigo and Razumova, 1973) was taken into consideration.

The soil gravimetric moisture was determined by drying the soil at 105 °C to a constant weight and measuring the soil sample weight before and after drying. Eq. 2 was used for calculating soil gravimetric water (Stadnic, 2010):

$$Soil \ moisture \ \% = \frac{\text{wet weight (g)} - \text{dry weight (g)}}{\text{dry weight (g)}} * 100 \quad \text{(Eq. 2)}$$

The humus content was determined according to the Tiurin method, based on the oxidation of soil humus in the solution by potassium dichromate and sulphuric acid, followed by the photocolourimetric determination of the trivalent chromium concentration (Mineev *et al.*, 2001). The mobile forms of phosphorus and exchangeable potassium were extracted with 0.5 M acetic acid and determined photocolourimetrically for phosphorous and photometrically for potassium as described by Sokolov and Askinazi (1965). Total N and P were determined by the Ginsburg method (Ginsburg, 1985) from the same soil samples.

4.2.5 Field layouts and application rates

Table 4.5 summarizes the treatment variants that were applied to the field. The nutrient requirement for corn of 60 kg of nitrogen per ha was based on local recommendations on the application of fertilizers for different crops (Andries *et al.*, 2013). The quantity of the fertilizer was calculated according to the amount of total nitrogen.

Table 4.5 Description of different fertilizer treatments applied to the corn field plots; data calculated from own measurements

Fertilizer applied	Total N kg tonne^{-1}	Total P kg tonne^{-1}	C content kg tonne^{-1}	Quantity of N applied kg ha^{-1}	Quantity of product applied kg ha^{-1}
Lactofermented mix and biochar	18.7	2.8	280	60	3200
Lactofermented mix	18.7	2.4	240	60	3200
Vermicomposted lacto-fermented mix and biochar (vermicast)	15.5	N/a	350*	80	5172
Mineral fertilizers (N12%, P52%), N$_{60}$P$_{60}$K$_{60}$ kg active ingredients ha^{-1}	12	5.2	-	60	115 NP fertilizer + 111.7 NH$_4$NO$_3$ and 150 KCl
Stored faeces	13.1	n.d	280	60	4580
Stored urine	9.6	n.d	-	60	6250
Stored cattle manure	17.8	n.d	260	60	3370

n.d - not determined.

The composition of the mixtures is given in Table 4.6. The application of all fertilizers was done manually, including the mineral fertilizer. Fertilization was carried out during the fall of both 2012 and 2013, except for the stored urine which was applied during spring, before sowing.

The experimental plots were 60 m^2 (5 x 12 m), distributed over a total experimental area of 1500 m^2. At harvesting, in order to avoid edge effects, one outer side row was omitted at each side. Moldroad ploughing was performed in the fall using a Belarusi 82 tractor, at 22 cm depth. Seeding of corn was done in April by a seeding-machine at 45000 plants/ha. Weeding, application of fertilizers and corn harvesting was done manually. The experimental field plots were positioned in a randomized block design with eight variants and three replications (Figure 4.3). The plots were fixed, so that each plot was treated twice. The plots where vermi-compost was applied were left unfertilized during the first year and fertilized only during the second year. All the applied fertilizers were distributed as uniformly as possible. The incorporation into the soil was done by tillage on the same day or the day after the application. Weed control was performed by hoeing two times during the entire growth season.

Table 4.6 The description of the composition of the fertilizers added

Fertilizer applied	Composition description
Lactofermented mix with biochar addition	Source separated faeces with sawdust as cover material (40 %), 6 month stored cattle manure (30 %), fruit waste from fresh juice boutiques (17%), molasses (8%), microbial inoculum from sauerkraut juice (3 %). After fermentation for 3 weeks, prior application, biochar preliminarily charged with nutrients from human urine was added.
Lacto-fermented mix	As above, but without biochar
Vermicast	Lacto-fermented mix which was vermicomposted for 4 months with *Eisenia foetida*
Mineral fertilizer (amofos)	As complex nitrogen and phosphorous NP mineral fertilizer, $N_{60}P_{60}K_{60}$ kg active ingredients ha^{-1}, N-12%, P 52%, supplemented by balanced amount of amonium nitrate (N34%) and potassium chloride (K40%).
Stored urine	6 months stored urine from UDDTs
Stored cattle manure	6 months cattle manure from a local farmer stored at ambient temperature.

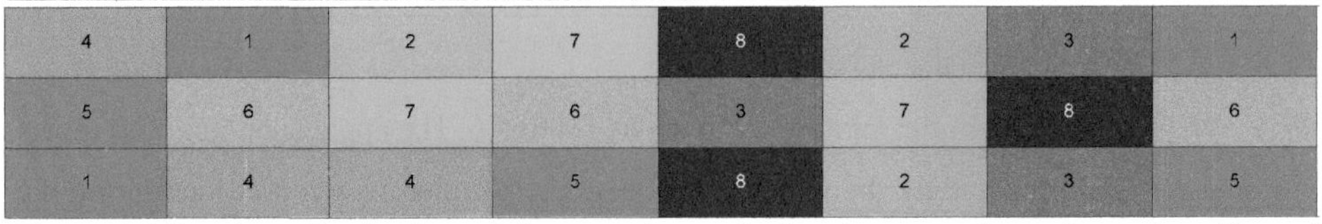

4	1	2	7	8	2	3	1
5	6	7	6	3	7	8	6
1	4	4	5	8	2	3	5

Figure 4.3 Random distribution of the experimental plots layout to investigate the effect of fertilizers on corn.

Fertilizers applied: 1-control with no fertilization; 2- mineral fertilizer (amofos); 3 - stored human faeces; 4 - stored cattle manure; 5 - lacto-fermented mix with no biochar; 6 - lacto-fermented mix with biochar; stored urine; vermicomposted lacto-femented mix and biochar.

4.2.6 Assessment of the effects of lacto-fermented mix and biochar on the growth, yield and yield components of corn

The corn was planted in April and harvested in September during the two consecutive years (2013 and 2014). Growth measurements were done at 7-10 leaves, during intensive growth (40 days from germination). These measurements were made on 10 randomly selected plants in each plot, for them the plant height (PH), stem diameter, the maximum leaf length (LL), and the maximum leaf width (LW) were recorded using a measuring tape (Figure 4.4) and a Vernier calliper.

The yield of corn was calculated by weighing the harvested corn cobs from each plot. Out of each sack, 10 randomly selected corn cobs were chosen and hand shelled; grain was weighed for evaluation of the ratio of grain to cob according to the requirements specified by the state standard for corn (GOST UJCN 13634-90, 2010). The following formula was used to calculate the yield (Dospehov, 1985):

$$Y = \frac{CG(100 - M)}{8600}$$

where Y - yield of corn at 14 % moisture (kg ha^{-1}), which is the standard storage corn moisture, according to the corn requirements for state purchases and deliveries (GOST UJCN 13634-90, 2010)

C - yield of cobs (weight of harvested corn cobs from each plot and estimated at kg/ha)

G - ratio of grain (%) (GOST UJCN 13634-90, 2010)

M - corn moisture (%)

8600 - coefficient of conversion of cob yield to grain yield at 14 % moisture.

Figure 4.4 Measuring the corn growth parameters in the field.

After the corn was harvested by standard methodology (Verhults *et al.*, 2013), ten corn ears were randomly chosen from the bags, for which the following yield component parameters were recorded: number of rows per ear, number of kernels per row, number of kernels per ear and average kernel weight. The number of rows per ear and the number of kernels per row were obtained by actual counting of the rows and kernels, while the number of kernels per ear was estimated by multiplying the number of rows per ear with the number of kernels per row. The average kernel weight per ear was obtained after hand shelling and weighing of kernels.

4.2.7 Statistical analysis

The experimental data were statistically processed via a Dunnet test of multiple comparisons within Anova one-way analysis of variance and using Minitab 17 statistical software. The means of all fertilizers were compared to the lacto-fermented mix with the biochar supplement for significant differences (at 95 % significance, based on p values and confidence interval CI).

4.3 Results

4.3.1 Influence of lacto-fermented mix and biochar on corn growth, yield and yield components

4.3.1.1 Growth of corn

Table 4.7 shows that application of a lacto-fermented mix supplemented by biochar during 2013 produced significantly taller plants compared to the unfertilized control and all other applied fertilizers ($p < 0.05$; CI contains only negative values) (Appendix A). The stem diameter was significantly bigger only in relation to the control. The leaf length was larger compared to the control and stored urine. No significant difference was observed in leaf width in comparison to the other applied fertilizers ($p > 0.05$ and CI contains both negative and positive values).

During 2014, the lacto-fermented mix supplemented by biochar again contributed to significantly taller plants, but this time only in relation to the control, stored faeces and vermicomposted lacto-fermented mix and biochar. The mineral fertilizer produced significantly taller plants than the lacto-fermented mix and biochar ($p > 0.05$, CI with both positive values). The stem diameter was not significantly different in comparison to the other fertilizers investigated. The leaf length was significantly longer only in comparison to the control and to the vermi-composted lacto-fermented mix and biochar, but significantly smaller than those fertilized by mineral fertilizers and stored faeces. The leaf width was significantly broader only in relation to the vermicomposted lacto-fermented mix and biochar.

Table 4.7 The effect of application of the lacto-fermented mix and biochar compared to other applied fertilizers on corn growth rate (mean ± SD, number of samples, n=3 per each plot) during 2013-2014.

Applied fertilizer	Plant height, cm	Stem diameter, cm	Leaf length, cm	Leaf width, cm
First production period, 2013				
Lacto-ferm mix and biochar	174.93 ± 1.01 A	2.67 ± 0.34 A	62.20 ± 1.80 A	9.66 ± 0.40 A
Lacto-ferm. mix only	160.56 ± 0.51*	2.53 ± 0.08 A	59.21 ± 1.71 A	9.20 ± 0.20 A
Mineral fertilizers	162.57 ± 1.16 *	2.31 ± 0.12 A	60.37 ± 1.27 A	9.07 ± 0.49 A
Stored faeces	157.03 ± 2.70 *	2.26 ± 0.21 A	59.43 ± 1.88 A	8.50 ± 0.61 A
Stored cattle manure	166.83 ± 1.02 *	2.41 ± 0.12 A	62.54 ± 1.35 A	8.97 ± 0.14 A
Stored urine	157.03 ± 2.70 *	2.22 ± 0.13 A	56.43 ± 1.19 *	7.88 ± 0.58 A
Control	145.37 ± 1.52 *	1.99 ± 0.22 *	56.29 ± 1.15 *	9.20 ± 1.57 A
Second production period, 2014				
Lacto-ferm mix and biochar	177.90 ± 0.80 A	2.67 ± 0.18 A	67.03 ± 1.42 A	9.53 ± 0.21 A
Lacto-fermented mix only	180.13 ± 1.03 A	2.62 ± 0.14 A	69.50 ± 0.66 A	8.95 ± 0.35 A
Mineral fertilizers	182.67 ± 0.35 **	2.75 ± 0.36 A	76.10 ± 0.85 **	9.48 ± 0.16 A
Stored faeces	170.73 ± 1.10 *	2.36 ± 0.01 A	72.33 ± 0.81 **	9.43 ± 0.13 A
Stored cattle manure	176.13 ± 0.81 A	2.43 ± 0.11 A	64.17 ± 1.26 A	9.15 ± 0.47 A
Stored urine	179.17 ± 1.26 A	2.56 ± 0.16 A	69.70 ± 4.45 A	8.84 ± 0.24 A
Control	168.00 ± 1.00 *	2.18 ± 0.15 A	59.97 ± 1.68 *	8.94 ± 0.37 A
Vermic lacto-ferm mix + biochar	174.07 ± 3.59 *	2.21 ± 0.48 A	61.20 ± 1.39 *	8.60 ± 0.36 *

The comparison was made by Dunnett multiple comparisons, where lacto-fermented mix and biochar was compared to all other fertilizers. Means labeled with letter A are not significantly different (p > 0.05, CI contains both positive and negative values), those not labeled with letter A are significantly different, those marked with * have significantly lower values p > 0.05 and CI contains both negative values), those with ** have significantly higher values (p < 0.05 and CI contains both positive values).

4.3.1.2 Grain yield and yield components

The lacto-fermented mix and biochar also contributed to a significantly higher grain yield during 2013 compared to all other applied fertilizers, with the exception of only the lacto-fermented mix, which showed no significant difference (Table 4.8, Appendix B).

Table 4.8 The effect of lacto-fermented mix and biochar compared to other applied fertilizers on corn yield during two different production periods (2013-2014) (mean ± SD, number of samples, n=3 per each plot).

APPLIED FERTILIZER	CORN YIELD, kg ha^{-1}
First production period	
Lacto-fermented mix and biochar	2368 ± 32 A
Lacto-fermented mix only	2352 ± 106 A
Mineral fertilizers	1556 *± 58
Stored faeces	2010 * ± 129
Stored cattle manure	2098 * ± 88
Stored urine	2015* ± 102
Control	1578 * ± 62
Second production period	
Lacto-fermented mix and biochar	3295 ± 118 A
Lacto-fermented mix only	2602 * ± 67
Mineral fertilizers	3404 ± 29 A
Stored faeces	2021 * ± 72
Stored cattle manure	2437* ± 118
Stored urine	2647* ± 67
Control	2136 * ± 72.4
Vermi-composted lacto-fermented mix supplemented by biochar	2416* ± 87

* - Parameters marked with a star have significantly lower value than the lacto-fermented mix and biochar.

The yield increase was in the following order compared to the unfertilized control: lacto-fermented mix and biochar > lacto-fermented mix > stored urine > stored faeces > stored cattle manure, which was 58 % > 45 % > 37 % > 32 % > 13 % > 9 % higher than the control. During 2014, the lacto-fermented mix and biochar again contributed to a significantly higher yield compared to the control and all other treatments, except for the mineral fertilizer, which was not significantly different (Table 4.7). The grain yield increased in the following order in relation to the control: mineral fertilizers > lacto-fermented mix and biochar > stored urine > lacto-fermented mix > stored cattle manure > vermi-composted lacto-fermented mix with a yield of 37 % > 35 % > 19 % > 18 % > 12 % > 12 % higher than the control. The lacto-fermented mix with biochar addition positively influenced the yield components of corn during both production years (Table 4.9).

Table 4.9 Influence of the lacto-fermented mix and biochar on yield components of corn in comparison to the unfertilized control and other types of fertilizers

1st Production year						
Yield component	Fertilizers applied	Means	Diff of levels	Diff of means	95 % CI	P-value
Nr rows ear[-1]	C	13.23 A	C - LFB	-0.367	(-1.996; 1.263)	0.964
	SU	12.93 A	SU -LFB	-0.667	(-2.296; 0.963)	0.698
	LF	13.9 A	LF - LFB	0.300	(-1.330; 1.963)	0.986
	SCM	12.63 A	SCM - LFB	-0.967	(-2.596; 0.663)	0.367
	MF	13.47 A	MF -LFB	-0.133	(-1.763; 1.496)	1.000
	SF	13.33 A	SF - LFB	-0.267	(-1.896; 1.363)	0.992
	LFB	13.6 A				
Nr kernels row[-1]	C	24.23	C - LFB	-7.53	(-12.54; -2.53)	0.003
	SU	27.10 A	SU -LFB	-4.67	(-9.67; 0.34)	0.072
	LF	32.73 A	LF - LFB	0.97	(-4.04; 5.97)	0.982
	SCM	24.33	SCM - LFB	-7.43	(-12.44; -2.43)	0.003
	MF	25.73	MF -LFB	-6.03	(-11.04; -1.03)	0.016
	SF	26.60	SF - LFB	-5.17	(-10.17; -0.16)	0.042
	LFB	31.77 A				
Nr kernels ear[-1]	C	327.07	C - LFB	-106.4	(-150.3; -62,5)	0.000
	SU	355.1	SU -LFB	-78.4	(-122.3; -34.5)	0.001
	LF	447.7 A	LF - LFB	14.2	(-29.7; 58.1)	0.850
	SCM	317.8	SCM - LFB	-115.7	(-159.6; -71.8)	0.000
	MF	346.2	MF -LFB	-87.3	(-131.2; -43.4)	0.000
	SF	352.3	SF - LFB	-81.2	(-125.1; -37.3)	0.000

	LFB	433.5 A				
Average kernel weight ear[-1]	C	119.9	C - LFB	-76.7	(-105.19; -8.21)	0.000
	SU	143.33	SU -LFB	-53.33	(-81.83; -4.84)	0.000
	LF	180 A	LF - LFB	-16.67	(-45.16; 11.83)	0.379
	SCM	121	SCM - LFB	-75.67	(-35.10; -9.23)	0.000
	MF	188.33 A	MF -LFB	-8.33	(-36.83 20.16)	0.895
	SF	153.33		-43.33	(-71.83; -14.84)	0.003
	LFB	196.7 A				

2nd Production Year						
Yield component	Variants	Means	Difference of levels	Difference of means	95% CI	P-value
Nr rows*ear[-1]	C	10.77	C - LFB	-2.900	(-4,517; -1.283)	0.001
	SU	12.47 A	SU -LFB	-1.200	(-2.817; 0.417)	0.188
	LF	11.83	LF - LFB	-1.833	(-3.450; -0,216)	0.024
	SCM	11.17	SCM - LFB	-2.500	(-4.117; -0.883)	0.002
	MF	14.00 A	MF -LFB	0.333	(-1.284; 1.950)	0.975
	SF	12.8 A	SF - LFB	-0.867	(-2.484; 0.750)	0.460
	LFB	13.67 A				
Nr kernels*row[-1]	C	28.47	C - LFB	-8.87	(-12.32; -5.41)	0.000
	SU	29.73	SU -LFB	-7.60	(-11.06; -4.14)	0.000
	LF	33.07	LF - LFB	-4.27	(-7.72; -0.81)	0.014
	SCM	24.33	SCM - LFB	-13.00	(-16.46; -9.54)	0.000
	MF	37.97 A	MF -LFB	0.63	(-2.82; 4.09)	0.986
	SF	30.63	SF - LFB	-6.70	(-10.17; -3.24)	0.000
	LFB	37.33 A				
Nr kernels*ear[-1]	C	315.8	C - LFB	-180.2	(-273.5; -86.9)	0.000
	SU	425.6 A	SU -LFB	-70.4	(-163.6; 22.9)	0.177
	LF	395.2	LF - LFB	-100.8	(-194.1; -7.6)	0.032
	SCM	317.8	SCM - LFB	-178.2	(-271,4 -84.9)	0.000
	MF	528.17 A	MF -LFB	32.2	(-61.1; 125.4)	0.815
	SF	407.7 A	SF - LFB	-88.3	(-181.6; 4.9)	0.066
	LFB	496 A				

Average kernel	C	119.97	C - LFB	-76.7	(-105.19; 8.21)	0.000
weight*ear^{-1}	SU	143.33	SU -LFB	-53.33	(-81.83; -24.84)	0.000
	LF	180.00 A	LF - LFB	-16.67	(-45.16; 11.83)	0.379
	SCM	121.00	SCM - LFB	-75.67	(-104.16; -7.17)	0.000
	MF	188.33 A	MF -LFB	-8.33	(-36.83; 20.16)	0.895
	SF	153.33	SF - LFB	-43.33	(-38.84; -17.16)	0.000
	LFB	196.7 A				

C- control; LFB - lacto-fermented mix and biochar; LF - lacto-fermented mix; SCM - stored cattle manure; MF- mineral fertilizer; SF - stored faeces. Means were compared by Dunnet multiple comparison, those not labeled with letter A are significantly different than LFB.

During the first year, the number of kernels per row, number of kernels per ear and average kernel weight per ear were significantly higher than the control, stored cattle manure, mineral fertilizer, stored faeces and stored urine ($p < 0.05$; CI with both negative values) (Table 4.9). The only exception came from the number of rows per ear, which was not significantly different than the other applied fertilizers and the control during the first production year ($p > 0.05$, confidence interval CI with both positive and negative values). During the second production year, the lacto-fermented faeces and bio-waste supplemented by biochar produced a higher number of rows per ear than the control, lacto-fermented mix only and stored cattle manure (Table 4.9). Mineral fertilizer produced significantly lower yield components and yield during the first year, but not significantly different yield and yield components during the second year (Table 4.9). Stored faeces yielded a significantly lower number of kernels per ear and average kernel weight per ear than the biochar supplemented lacto-fermented faeces and bio-waste mix.

Addition of biochar had an added value to the yield components. The effect was observed during the second year only when the lacto-fermented mix without biochar had a significantly lower number of rows per ear, kernels per row and kernels per ear. It, however, did not have any influence on the average kernel weight (Table 4.9). As in the case of yield components, the yield of corn plots fertilized by stored cattle manure, faeces and urine as well as the unfertilized control had significantly lower yields than the plots fertilized by lacto-fermented faeces and bio-waste, supplemented by biochar during both production years (Figure 4.5).

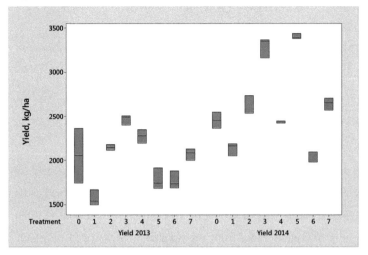

Figure 4.5 Yield of corn at 14 % moisture (kg ha^{-1}) during the 1st production period (2013) and 2nd production period (2014).
1- Control; 2- stored urine; 3- lacto-fermented mix and biochar; 4 - lacto-fermented mix;
5 - stored cattle manure; 6 - mineral fertilizer; 7 - stored faeces

The variations on ear length and grain filling are shown in Figure 4.6, the results
presented are from the second production year.

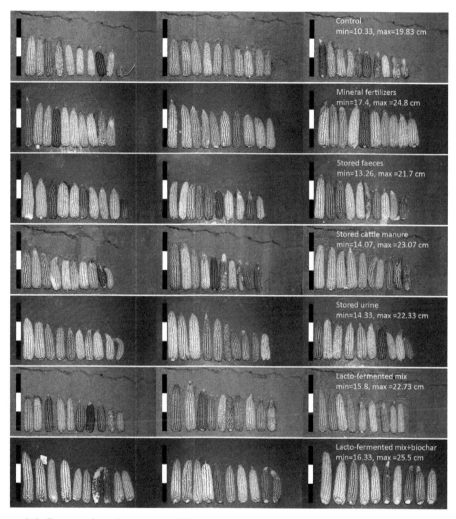

Figure 4.6. Corn ear length and grain filling in corn fertilized by the lacto-fermented mix of
faeces and bio-waste supplemented by biochar in comparison to other applied fertilizers (each
column represents one of the triplicates of the applied fertilizers).

The corn fertilized by the lacto-fermented mix of faeces and bio-waste, supplemented by biochar as well as mineral fertilizer produced the longest ears (Figure 4.6) compared to the control and other fertilizing mixes. The shortest ears and the highest number of ears with reduced number of kernels were observed in the unfertilized control. A large number of corn plants with short ears were also observed in corn fertilized by the stored faeces, cattle manure and urine alone, compared to the investigated mix of lacto-fermented faeces and bio-waste supplemented by biochar. During the first production year, the yield of all applied fertilizers was significantly lower than the average yield for Moldova during 2013 (Table 4.10).

Table 4.10 Corn yield of the applied fertilizers in comparison to the average value (based on spring and summer precipitation data) and lacto-fermented mix supplemented by biochar

Fertilizers applied	Corn yield kg ha^{-1} (mean ± SD)[1]	Difference of yield of applied fertilizers compared to the average yield of Moldova during 2013 and 2014
	1st production year	
Control	1578 ± 62	-49 *
Stored urine	2015 ± 30	-35 *
Lacto-fermented mix	2352 ± 106	-24 *
Stored cattle manure	2098 ± 88	-32 *
Mineral fertilizer	1556 ± 58	-50 *
Stored faeces	2010 ± 129	-35 *
Lacto-fermented mix and biochar	2368 ± 32	-23 *
	2nd production year	
Control	2136 ± 87	-36 *
Stored urine	2647 ± 67	-21 *
Lacto-fermented mix	2602 ± 67	-22 *
Stored cattle manure	2098 ± 88	-37 *
Mineral fertilizer	3404 ± 29	+2
Stored faeces	2021 ± 129	-40 *
Lacto-fermented mix and biochar	3295 ± 118	-2

*significantly lower, no mark means not significantly different.

During the second production year, only lacto-fermented faeces and bio-waste, supplemented by biochar as well as the mineral fertilizer produced yields that are not significantly different from the average yield for Moldova during 2014 (Table 4.10). It should be noted that two years of application is a rather restricted period to draw conclusions on the effect of fertilizer use efficiency on yield in comparison to the overall corn yield at the country level.

4.3.2 Influence of the lacto-fermented mix and biochar on soil quality

4.3.2.1 Soil available water storage capacity

During the first production year, fertilization by the lacto-fermented mix and biochar had contributed to a significantly higher soil available water storage capacity in the soil layer 0-20 cm compared to the control. This value was, however, significantly lower than the fertilization by cattle manure. During the second production year, no significant difference of the available water storage capacity in the soil layer 0-20 cm was observed in comparison to the control and the other applied fertilizers. In the deeper soil layer of 20-40 cm, stored cattle manure contributed again with a significantly higher soil available water storage capacity (Table 4.11), whose values (26-86 mm) were in conformity with the norm of productive moisture stocks for soil required for corn (Bugaeva and Mironova, 2009). Cattle manure increased the humus content during both production years and both soil layers (Table 4.12). In spite of the increasing soil available water storage capacity, cattle manure produced a significantly lower yield than lacto-fermented faeces and bio-waste, supplemented by biochar, probably due to a higher loss of nitrogen as NH_3 via volatilization or via leaching.

Data on average precipitation during spring and summer for the first and second production periods (Table 4.1) show that for both spring and summer, the precipitation was significantly higher than the average precipitation for the Nisporeni district (Nedealkov *et al.*, 2013). Apparently, the first production period received a more favourable amount of precipitation in spring and summer compared to the second production period. However, there was a much more uniform distribution of the rainfall during the key corn growing periods of the second production year (i.e. June and July). This is likely to have positively influenced the yield. During both production years (Table 4.2), the seasonal rainfall was less than 81 mm, a below average value during summer or spring. Such conditions may lead to a corn yield decreased by 15% (Vronskih, 2014).

4.3.2.2 Soil available nitrate content

The C:N ratio in the lacto-fermented mix supplemented by biochar, lacto-fermented mix alone and cattle manure was < 15 (Andreev *et al.*, 2016). Such a C:N ratio should have accelerated the organic matter decomposition and increased the availability of mineral nitrogen (Qian and Schoenau, 2002); however, probably higher losses of N have occurred after application of cattle manure. In the case of stored faeces, which contained sawdust, added during the collection stage, might not have been fully decomposed, the C:N ratio was > 15 (Andreev *et al.*, 2016). With this range of C:N ratio, part of the nitrogen from the applied stored faeces was probably immobilized by microbial consortia that continued to decompose the material into the soil (Qian and Schoenau, 2002).

The mineral fertilizer gave the highest nitrate content in both soil layers during the first production year. The lacto-fermented mix of faeces and bio-waste, supplemented by biochar, has increased significantly the soil nitrate concentration in the 0-20 cm soil layer during the first production year, but only compared to the control and stored urine (Table 4.11). During the second year, this particular fertilizer has increased significantly the soil nitrate content in relation to the control, stored urine and faeces only in the 20-40 cm soil layer, while for the 0-20 cm soil layer there was no significant difference compared to all fertilizers applied (Table 4.11).

Table 4.11 Influence of the lacto-fermented mix and biochar on soil moisture and nitrate content in comparison to the unfertilized control and other types of fertilizers

Soil compon.	1st Production year					
	Variant	Mean	Difference of levels	Diff of means	95% CI	P-value
Soil available water storage capacity, mm, 0-20 cm	C	20.9	C - LFB	-5.48	(-10.37; -0.58)	0.026
	SU	28.8 A	SU -LFB	2.38	(-2.51; 7.28)	0.549
	LF	27.5 A	LF - LFB	1.11	(-3.79; 6.01)	0.962
	SCM	25.5 A	SCM - LFB	-0.89	(-5.79; 4.00)	0.367
	MF	28.9 A	MF -LFB	2.49	(-2.40; 7.39)	0.507
	SF	28.0 A	SF - LFB	1.59	(-3.30; 6.49)	0.847
	LFB	26.4 A				
	C	64.4 A	C - LFB	8.19	(-16.50; 32.87)	0.837
	SU	29.2	SU -LFB	-27.00	(-51.68; -2.31)	0.030
	LF	64.3 A	LF - LFB	8.07	(-16.61; 32.75)	0.844
	SCM	86.2	SCM - LFB	29.95	(-12.44; -2.43)	0.021
	MF	61.1 A	MF -LFB	4.88	(2.63; 35.93)	0.016
	SF	71.6 A	SF - LFB	15.35	(-3.38; 29.91)	0.063
	LFB	56.2 A				
Soil nitrate content, kg ha⁻¹ 0-20 cm	C	39.9	C - LFB	-21.40	(-33.52; -9.28)	0.001
	SU	38.9	SU -LFB	-22.43	(-34.55;-10.31)	0.000
	LF	48.8 A	LF - LFB	-12.47	(-24.59; -0.35)	0.043
	SCM	67.7 A	SCM - LFB	6.40	(-5.72; 18.52)	0.473
	MF	81.9	MF -LFB	20.60	(8.48; 32.72)	0.001
	SF	53.0 A	SF - LFB	-8.27	(-20.39; 3.85)	0.249
	LFB	61.3 A				
Soil nitrate content, kg ha⁻¹ 20-40 cm	C	115.1 A	C - LFB	-28.5	(-62.2; 5.1)	0.112
	SU	173.2 A	SU -LFB	29.6	(-4.1; 28.8)	0.095
	LF	138.8 A	LF - LFB	-4.8	(-38.5; 11.83)	0.996
	SCM	160.6 A	SCM - LFB	17.0	(-16.7; 50.7)	0.515
	MF	191.6	MF -LFB	-26.4	(14.4; 81.7)	0.005
	SF	117.2 A	SF - LFB	-26.4	(-60.1; 7.3)	0.153
	LFB	143.3 A				

			2nd Production Year			
Soil component	Variant	Mean, mm	Difererence of levels	Differ. of means	95% CI	P-value
Soil	C	29.2 A	C - LFB	-5.64	(-12.28; 1.00)	0.111
available	SU	26.9 A	SU -LFB A	-2.28	(-8.92; 4.36)	0.817
water	LF	26.1 A	LF - LFB A	-3.05	(-9.69; 3.59)	0.600
storage	SCM	27.7 A	SCM - LFB A	-1.52	(-8.16; 5.12)	0.961
capacity,	MF	30.71 A	MF -LFB A	1.52	(-5.12; 8.16)	0.961
mm, 0-20	SF	29.2 A	SF - LFB A	0.01	(-6.63; 6.65)	1.000
cm	LFB	29.2A				
Soil	C	59.8 A	C - LFB	1.72	(-14.92; 18.37)	0.999
productive	SU	72.3 A	SU -LFB	14.19	(-2.45; 30.84)	0.109
moisture,	LF	55.4 A	LF - LFB	-2.69	(-19.34; 13.95)	0.992
mm, 20-40	SCM	77.3	SCM - LFB	19.28	(2.63; 35.93)	0.021
cm	MF	74.0 A	MF -LFB	15.92	(-0.72; 32.57)	0.063
	SF	71.3 A	SF - LFB	13.27	(-3.38; 29.91)	0.144
	LFB	58.0 A				
Soil nitrate	C	60 A	C - LFB	-19.55	(-44.02; 4.93)	0.143
content, kg	SU	73 A	SU -LFB	-22.30	(-46.77; 2.17)	0.177
ha[-1], 0-20	LF	72.4 A	LF - LFB	1.71	(-22.76; 6.19)	1.000
cm	SCM	77.2 A	SCM - LFB	6.50	(-17.97; 0.97)	0.927
	MF	78 A	MF -LFB	-7.21	(-31.68; 7.27)	0.89
	SF	49.5 A	SF - LFB	-21.16	(-45.63; 3.31)	0.102
	LFB	58.0 A				
Soil nitrate	C	119.9	C - LFB	-73.1	(-115.7; -30.4)	0.001
content, kg	SU	142.1	SU -LFB	-50.8	(-93.4; -8.2)	0.017
ha[-1], 20-40	LF	190.3 A	LF - LFB	-2.6	(-45.2; 40.0)	1.000
cm	SCM	210.2 A	SCM - LFB	17.2	(-25.4; 59.9)	0.706
	MF	187.9 A	MF -LFB	-5.0	(-47.6; 37.6)	0.999
	SF	112.4	SF - LFB	-80.6	(-126.2; -37.9)	0.000
	LFB	192.9 A				

The values labeled with letter A are not significantly different than lacto-fermented mix and biochar. C - control, SU - stored urine, LF - lacto-fermented mix, without biochar, SCM - stored cattle manure, MF - mineral fertilizer, SF - stored faeces, LFB - lacto-fermented mix and biochar.

Under such dry periods, corn intensively consumes nitrogen and water for vegetative growth and reproductive development. Under the conditions of reduced precipitation, the analysed spring soil moisture may not fully reflect the complete influence of the applied fertilizers on soil moisture. In spite of the increasing soil available water storage capacity, cattle manure produced a significantly lower yield than the lacto-fermented mix and biochar.

The mineral fertilizer contributed to the highest nitrate content in both soil layers during the first production year. The lacto-fermented mix of faeces and bio-waste, supplemented by biochar increased significantly the soil nitrate concentration in the soil layer 0-20 cm during the first production year, but only compared to the control and stored urine (Table 4.11). During the second year, this particular fertilizer has increased significantly the soil nitrate content in relation to the control, stored urine and faeces only in the soil layer 20-40 cm, while for the 0-20 cm soil layer there was no significant difference compared to all applied fertilizers (Table 4.11).

4.3.2.3 Mobile phosphorus and exchangeable potassium

The content of mobile phosphorous and exchangeable potassium of the investigated soil refers to a medium nutrient content (Sokolov and Askinazi, 1965). During the first year of application, the available phosphorus in both soil layers did not differ significantly in comparison to the other applied fertilizers and the unfertilized control in any of the production years investigated. An accumulation of mobile phosphorus and exchangeable potassium was observed during the second production year for all treatments compared to the control (Table 4.12, Appendices D and E). The lacto-fermented mix supplemented by biochar contributed to significantly higher values of phosphorus in the soil layer 0-20 cm compared to the control, stored cattle manure and the vermi-composted lacto-fermented mix supplemented by biochar. For the deeper soil layer (20-40 cm), significantly higher phosphorus values were found only in relation to the control. The values were, however, significantly lower than in the plots fertilized by the stored urine. The potassium content increased significantly during the first production year after application of the lacto-fermented mix supplemented by biochar, but only in the soil layer 20-40 cm (Table 4.12).

Table 4.12 Mobile P[1] and exchangeable K[2] content in the soil with different fertilizer treatments during 2013-2014 (number of samples, n = 3 per each plot)

Applied fertilizer	P_2O_5 (mg kg^{-1})		K_2O (mg kg^{-1})	
First production period	**0-20 cm**	**20-40 cm**	**0-20 cm**	**20-40 cm**
Lacto-fermented mix and biochar	110.6 ± 4.6 A	101.7 ± 4.8 A	72.0 ± 2.5 A	75.3 ± 3.0 A
Lacto-fermented mix	103.0 ± 6.7 A	101.0 ± 2.9 A	80.0 ± 1.6 **	64.5 ± 1.8 *
Mineral fertilizers	108.8 ± 2.1 A	105.2 ± 5.9 A	73.3 ± 1.4 A	78.5 ± 1.9 A
Stored cattle manure	107.8 ± 10.6 A	102.3 ± 5.9 A	68.1 ± 2.32 A	62.5 ± 1.3 *
Stored faeces	105.6 ± 4.9 A	96.7 ± 13.3 A	69.7 ± 0.87 A	63.6 ± 1.9 *
Stored urine	107.8 ± 10.6 A	101.9 ± 6.9 A	75.0 ± 2.5 A	64.1 ± 1.9 *
Control	99.9 ± 3.4 A	103.0 ± 6.4 A	69.1 ± 0.8 A	61.7 ± 1.3 *
Second production period	**0-20 cm**	**20-40 cm**	**0-20 cm**	**20-40 cm**
Lacto-fermented mix and biochar	132.6 ± 12.7 A	113.0 ± 3.6 A	81.8 ± 5.5 A	78.6 ± 22.7 A
Lacto-fermented mix	127.9 ± 8.9 A	116.4 ± 2.3 A	80.7 ± 4 A	83.0 ± 7.6 **
Mineral fertilizers	137.8 ± 1.4 A	110.5 ± 2.9 A	100.7 ± 7.8 A	105.1 ± 4.6 **
Stored faeces	139.7 ± 0.6 A	118.7 ± 2.8 A	92.4 ± 1.2 A	90.7 ± 1.8 **
Stored cattle manure	112.2 ± 3.0 *	114.5 ± 1.03 A	96.1 ± 7.6 A	98.4 ± 7.6 **
Stored urine	134.4 ± 0.5 A	130.2 ± 4.10 **	81.4 ± 3.3 A	97.3 ± 22.7 **
Control	101.3 ± 3.1 *	99.5 ± 0.00 *	78.7 ± 0.2 A	65.5 ± 0.00 A
Vermi-composted lacto-fermented mix and biochar	117.2 ± 1.6 *	111.2±8.0 A	83.0±7.6 A	67.6 ± 3.56 A

The values labeled with letter A are not significantly different than lacto-fermented mix and biochar. *values marked with one star are significantly lower than the lacto-fermented mix and biochar and those marked with two stars ** are significantly higher than lacto-fermented mix and biochar. [1,2] - expressed as K and P respectively, see Appendix F.

In the soil layer 0-20 cm, a significant increase was observed only for the lacto-fermented mix without biochar. During the second production year, in the soil layer of 0-20 cm, there was no significant difference of the potassium content in comparison to the other fertilizers investigated. Nevertheless, in the soil layer 20-40 cm, the potassium content was significantly lower compared to the other applied fertilizers, except for the control, where no significant difference was found.

4.3.2.4 Humus content

No significant difference in the humus content was observed after fertilization with the
lacto-fermented mix supplemented by biochar in comparison to the control plots and other
fertilizers applied in either of the two production years investigated (Table 4.13, Appendix C).

**Table 4.13 Humus content in the soil layers 0-20, 20-40 cm during the first and second
production years (number of samples, n = 3 per each plot)**

Applied fertilizer	Humus content, %	
	0-20 cm	20-40 cm
First production period		
Lacto-fermented mix and biochar	2.80 ± 0.09 A	1.92 ± 0.08 A
Lacto-fermented mix only	2.98 ± 0.20 A	1.92 ± 0.05 A
Mineral fertilizers	2.59 ± 0.11 A	1.57 ± 0.04 A
Stored faeces	2.59 ± 0.38 A	1.90 ± 0.12 A
Stored cattle manure	2.81 ± 0.15 A	2.16 ± 0.28 A
Stored urine	2.57 ± 0.07 A	1.81 ± 0.05 A
Control	2.69 ± 0.18 A	1.91 ± 0.18 A
Second production period		
Lacto-fermented mix and biochar	2.68 ± 0.08 A	2.71 ± 0.13 A
Lacto-fermented mix only	2.76 ± 0.12 A	2.79 ± 0.12 A
Vermicomposted lacto-fermented mix and biochar	2.75 ± 0.37 A	2.64 ± 0.32 A
Mineral fertilizers	2.46 ± 0.35 A	2.39 ± 0.47 A
Stored faeces	2.53 ± 0.37 A	2.33 ± 0.35 A
Stored cattle manure	2.85 ± 0.06 A	2.81 ± 0.06 A
Stored urine	2.48 ± 0.17 A	2.42 ± 0.16 A
Control	2.54 ± 0.32 A	2.44 ± 0.29 A

The values labeled with letter A are not significantly different from the lacto-fermented mix and
biochar.

For the layer 20-40 cm in the first year, the data showed a big difference with the
humus content compared to that of the second production year. This might have been
influenced by a different degree of oxidation of the organic material during laboratory

analysis or non-uniformity of organic material incorporation in the soil. The total amount of carbon applied with the lacto-fermented mix supplemented by biochar and lacto-fermented mix without biochar was lower compared to the stored faeces and vermi-composted lacto-fermented mix supplemented by biochar. However, a higher (but not significantly higher) humus content was encountered for these fertilizers during the first production year. A slight decrease in the humus content was, nevertheless, observed during the second year (Table 4.12).

4.3.2.4 Bulk density

During the first production year, the lacto-fermented mix supplemented by biochar has contributed to a significantly lower bulk density in the superior layer of the soil (0-20 cm) compared to the mineral fertilizer and stored faeces (Table 4.14, Appendix F).

Table 4.14 Bulk density (g/cm^3) in the soil layers 0-20 and 20-40 cm during spring of 2013 and 2014 (number of samples, n = 3 per each plot)

Applied fertilizer	1st production period		2nd production period	
	0-20 cm	20-40 cm	0-20 cm	20-40 cm
Lacto-fermented mix and biochar	1.05±0.05 A	1.28±0.02 A	1.04±0.03 A	1.27±0.02 A
Lacto-fermented mix	1.14±0.04 A	1.23±0.03 A	1.08±0.03 A	1.25±0.03 A
Mineral fertilizer	1.38±0.02 **	1.40±0.05 A	1.23±0.01 **	1.47±0.07 **
Stored faeces	1.23±0.02 **	1.27±0.03 A	1.07±0.02 A	1.41±0.05 **
Stored cattle manure	1.14±0.03 A	1.13±0.03 *	1.14±0.06 A	1.41±0.03 **
Stored urine	1.11±0.03 A	1.27±0.06 A	1.21±0.07 **	1.42±0.06 **
Control	1.14±0.08 A	1.25±0.02 A	1.08±0.03 A	1.33±0.07 A
Lacto-feremented mix and biochar +V[1]			1.04±0.03 A	1.27±0.02 A

The values labeled with letter A are not significantly different than the lacto-fermented mix and biochar. *values marked with one star are significantly lower than the lacto-fermented mix and biochar and those marked with two stars ** are significantly higher than the lacto-fermented mix and biochar. [1] lacto-fermented mix and biochar was supplementary vermi-composted.

For the deeper soil layer (20-40 cm), no significant difference in the bulk density was found compared to other fertilizers, except for cattle manure which had a significantly lower bulk density. During 2014, similar to 2013, a significantly lower bulk density was observed after application of the lacto-fermented mix supplemented by biochar compared to mineral fertilizer and stored urine. For the deeper soil layer (20-40 cm), the bulk density was significantly lower than in the mineral fertilizer, stored cattle manure, stored faeces and urine. The application of mineral fertilizer (layers 0-20 and 20-40 during 2013 and layer 20-40 cm during 2014) contributed to a bulk density exceeding 1.35 g/cm^3 (Table 4.14), such values might show a tendency of soil compaction (Andries *et al.*, 2014). The same tendency was also observed during 2014 in the 20-40 cm soil layer for the plots fertilized by cattle manure, stored faeces, and urine.

4.4 Discussion

4.4.1 Corn growth

This study demonstrated that the lacto-fermented mix and biochar had a beneficial effect on the growth of corn. Although some of the growth parameters were not consistent for both production years, a significant increase in the key growth parameters such as corn height and leaf length was observed. In spite of the fact that during 2014, the mineral fertilizer had significantly better effects on the corn height and leaf length than the lacto-fermented mix supplemented by biochar, it did not have a significantly higher yield during that specific year (Tables 4.7 and 4.8).

4.4.2 Corn yield and yield components

The corn yield fertilized by the lacto-fermented mix supplemented by biochar was very similar to the average for Moldova (Table 4.10) during 2006-2013 (National Bureau of Statistics of the Republic of Moldova, 2014). Considering the low humus content of the soil investigated and also the lack of application of any fertilizers to the area, a beneficial effect from the fertilization with the lacto-fermented mix and biochar was observed even at a relatively low application rate (60 kg N active ingredients ha^{-1}) for a crop like corn, which requires a high nitrogen fertilization rate, e.g. 120-180 kg ha^{-1} (Andries, 2007).

During the two year field application, lacto-fermented faeces and bio-waste, supplemented by biochar had also an overall significantly higher effect on yield and yield components of corn than the unfertilized control, lacto-fermented mix of faeces and bio-waste without biochar, mineral fertilizer, stored cattle manure, faeces and urine (Table 4.9). The corn plants had bigger kernels, longer ears, and fewer ears with reduced numbers of kernels (Figure 4.6). The organic matter from the lacto-fermented faeces and bio-waste, which was stabilized by the supplemented biochar and the biochar properties themselves (Andreev *et al.*, 2016) might have provided a more efficient use of moisture during flowering and grain production. Organic matter adds to an improvement of pore volume and water conductivity as well as water retention (Fischer and Glaser, 2012). Biochar also improves soil physical properties such as aggregate stability, pore size distribution and, therefore, water use efficiency (Obia *et al.*, 2016). Such conditions probably allowed the corn fertilized by the lacto-fermented mix of faeces and bio-waste, supplemented by urine charged biochar to develop a better root system than the other applied fertilizers and a subsequent more intensive growth (Andreev *et al.*, 2016), even during the conditions of reduced precipitation, and ultimately produced a higher yield. Also the yield was enhanced as the investigated fertilizing mix has beneficially affected several soil parameters, such as soil potassium and humus content as well as soil bulk density (Andreev *et al.*, 2016).

4.4.3 Effect of precipitation on yield and yield components

The corn fertilized by the mineral fertilizer had an increased yield compared to the control and all other fertilizers only during favourable precipitation conditions, while the lacto-fermented mix supplemented by biochar favoured a yield increase during both years, regardless of the weather conditions (Table 4.8). This is probably related to the capacity of the fertilizer, especially its biochar component, to retain nitrates from leaching and potentially improving the soil water holding capacity (Kammann *et al.*, 2015).

The difference in yield during the first and second years of the application of mineral fertilizer can be explained by variations in intensity and quantity of annual precipitation and temperature during 2013 and 2014 (Table 4.1), that might have influenced nutrient availability and leaching (Udawatta *et a*l., 2006). Indeed, Table 4.1 shows that during the first production period, there was a wetter fall season which might have contributed to accumulation of soil moisture stocks during spring. Also, abundant precipitation during April and May, twice as much as the previous year, occurred during the second production period (Figure 4.7).

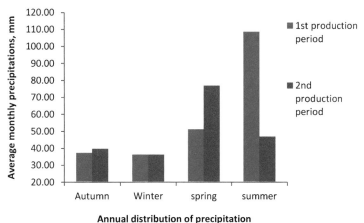

Figure 4.7 Difference in the seasonal distribution of precipitation during first and second production years.

In addition for this second production year, the whole month of July had a uniform precipitation distribution, with a maximum precipitation after 20 of July (Anonymous, 2015), the period of corn intensive growth and use of soil moisture and nitrogen. This probably favoured both more efficient mineral nitrogen uptake and intense accumulation by the stem and leaves in the plots fertilized by the lacto-fermented mix of faeces and bio-waste, supplemented by biochar. Moreover, such conditions during the second production year most probably also allowed a more efficient translocation from the vegetative parts to the grain during the dry spells of August and September than during the first production year.

In contrast, during the first production year, heavy rains were encountered during the end of June and beginning of July, while the rest of the month was dry with high temperatures encountered. The thunderstorms during June-July (Anonymous, 2015) coupled with a wetter autumn-winter-early spring period during 2012 (Table 4.1), possibly contributed to nitrate losses via leaching from the plots fertilized by mineral fertilizer, stored cattle manure, stored faeces and urine, thus negatively affecting the pre-anthesis period and kernel setting (Runge, 1968). In addition, the temperature at the end of June and beginning of July was more favourable during 2014 than 2013, thus positively influencing the grain yield.

Application of such fertilizers to crops with a high soil fertility requirement as corn (Paponov and Engels, 2005) can thus be an alternative to mineral fertilizer, stored cattle manure, faeces or urine alone, particularly under the conditions of decreased organic matter content, frequent drought or heavy rains as those occurring in Moldova (Potop and Soukup, 2009; Potop, 2011). Therefore, building up of the soil organic matter through the application of lacto-fermented faeces and bio-waste in combination with biochar could be more favourable than the application of mineral fertilizers alone.

More research is needed on the corn nitrogen uptake and nitrogen distribution within different vegetative parts of the corn at anthesis and grain filling. There is a big difference in the nitrogen uptake at the different development stages of corn (Spiertz and De Vos, 1983). During the growing period, nitrogen uptake exceeds that of mineralization and the nitrogen reserves in the soil are very quickly depleted. During grain filling (e.g. August-September), the nitrogen supply from the soil becomes less important, as nitrogen is being relocated from the vegetative parts (Spiertz and De Vos, 1983). Additional knowledge is required on the proportion of nitrogen accumulated in the beginning of the growing period and that translocated from vegetative parts to the grain during the grain filling period. Such research will offer information on the amount of nitrogen to be applied to the soil with a lacto-fermented mix of faeces and bio-waste, supplemented by biochar and the timing of the fertilizer application so as to maximize grain yield.

4.4.4 Effect of soil available water and nitrate content on yield components

An increase in the soil available water storage capacity might be related to an enhancement in the soil organic matter, which raised the water holding capacity of the soil (Diacono and Montemurro, 2010). The lacto-fermented mix of faeces and bio-waste supplemented with biochar or without biochar have both contributed to a high humus content during the first year, which slightly decreased during the second year (Andreev *et al.*, 2016). Additional research is important on the changes induced by the lacto-fermented mix supplemented by biochar on the soil available water storage capacity along the whole soil profile (0 to 100 cm) and for the whole corn development stage.

The average precipitation during spring of the first production year was significantly higher than the average multiannual precipitation for the Nisporeni district (Nedealcov *et al.*, 2013). However, the precipitation value during the studied period was still far below the average rainfall for Moldova, such an amount of spring precipitation is more close to the conditions characterized by a dry spring (Vronskih, 2014). Under dry conditions, nitrate accumulation in the soil may take place (Lucci *et al.*, 2013). During June and the first half of July, corn requires a high soil nitrogen content, since it is the period of intensive vegetative growth, when the root system, corn panicle and primordia of the ears are formed. However, spring dry spells during the first production year might have decreased the plant capacity to take up nitrogen (Lucci *et al.*, 2013). Rapid rewetting, for example during flushes from heavy rains after the dry period may cause intense leaching of nitrates. Therefore, at the end of June, in the first production year, 51% of all monthly precipitation has fallen during 2 days only and in the beginning of July 69% during one day only (Anonymous, 2015). Studies have shown that during June, up to 20 to 76% of NO_3^- leaching may occur (Corrêa *et al.*, 2005).

Another period which might have also contributed to a decrease in the corn yield components and yield for the unfertilized control or corn fertilized by mineral fertilizer, stored faeces, urine and cattle manure during the first production year, is the second half of July and the whole month of August to beginning of September. During this time, corn has a high water and nitrogen demand for internode elongation, tasseling, silking, pollination, kernel setting and filling (Pandrea, 2012). The two weeks following pollination up to maturity (October) are crucial in yield formation as the final kernel weight is determined in that period. Thus, the soil moisture stress or nitrogen deficiency during this period of time may result in reduced kernel size and consequently in reduced yields (Hollinger and Changnon, 1991) as in our case happened for the unfertilized control and plots fertilized by stored faeces, cattle manure or urine alone (Table 4.9, Figure 4.5).

Lack of nitrogen during June might have contributed to ovule abortion, thus resulting in fewer kernels (Gallais and Hirel, 2004), which was indeed observed in Figure 4.6. Nitrate is a mobile anion, not absorbed by soil organic matter particles. It is highly available following application of mineral fertilizer or biosolids and can thus easily be leached below the root zone, leading to groundwater pollution (FAO, 1996). For example, in a study on pheosem soil (Esteller *et al.*, 2009), among the applied fertilizers, mineral fertilizer had the highest proportion of nitrate leached (37 % and 24 % during the first and second year, respectively), followed by composted (17 % and 14 %) and untreated (11 %) biosolids. Under the conditions of insufficient soil nitrogen as those caused by nitrate leaching as a result of heavy summer rains, corn can remobilize nitrogen from other parts such as the stem or leaves. However, the dry period which continued over August and early September (Anonymous, 2015), coupled with high temperatures, might have caused leaf senescence and thus loss of accumulated nitrogen. As a consequence, a decrease in the number of filled kernels leading to an overall yield decrease might have occurred in the plots on which mineral fertilizer, stored cattle manure, stored faeces or urine were applied.

An integrated view on the efficiency of nitrogen use from the lacto-fermented mix of faeces and bio-waste, supplemented by biochar could have been obtained if all mineral nitrogen species would have been measured, including $N-NO_3^-$ and NH_4^+-N. Field experiments could identify the level of mineralization or immobilization of nitrogen in the applied fertilizers. In such experiments, soil can be placed in polyethylene bags and buried in the 0-60 cm horizon and the mineral nitrogen content in the bag can be compared to that in the root zone at selected time intervals. The bags are retrieved and the nitrogen content measured (Westermann and Crothers,

1980; Lentz *et al.*, 2012), the net N mineralization being calculated by subtracting the inorganic N concentration in the initial soil from that in the soil of the retrieved bag.

4.4.5 Humus content

Although mineral fertilizers contribute to the availability of phosphorus and exchangeable potassium in the soil profile, a loss of soil organic carbon may also occur. For example, long term experimental research carried out on different crops on chernoziom soils demonstrated a carbon loss of up to 28-34% in the soil profile of 0-40 cm and 66-72 % in the 40-100 cm profile in the plots treated with mineral fertilizers compared to the unfertilized control (Boincean *et al.*, 2014). In the absence of a supplementary carbon source, mineral fertilizers may lead to intensification of humus decomposition and degradation of soil fertility. At the same time, even though the lowest amounts of organic matter were added with the lacto-fermented mix supplemented by biochar as well as the lacto-fermented mix without biochar, these fertilizers contributed to the highest humus content during the first year. The slight decrease observed during the second production year may be the cause of the priming effect, an increase in the decomposition of soil organic matter due high microbial activity after application of fresh organic matter. Incubation experiments or C-labelling in the applied substrates would help to trace the decomposition of the added organic matter and the portion that remains stable in the soil and its effect on mineralization of the native soil organic matter (Perelo and Munch, 2005).

In the first production year, there were variations in the humus content in the 20-40 cm layer compared to the second year (Table 4.11). Such differences might be caused by a non-uniform distribution of organic matter (during application or ploughing). Two years is a rather short period for the evaluation of the changes in the soil organic matter content and long time research (e.g. 5-7 years) is required to validate these findings.

4.4.6 Potassium content

The potassium content increased significantly during the first production year after application of the lacto-fermented mix supplemented by biochar, but only in the soil layer

20-40 cm. In the soil layer 0-20 cm, a significant increase was observed only for the lacto-fermented mix without biochar. During the second production year, in the soil layer of 0-20 cm, there was no significant difference in the potassium content in comparison to the other investigated fertilizers. Nevertheless, in the soil layer 20-40 cm, the potassium content was significantly lower compared to the other applied fertilizers, except for the control, where no significant difference was found.

The lacto-fermented mix supplemented by biochar also had a positive influence on the soil mobile potassium content during the first year of application (soil layer 20-40 cm) compared to all applied fertilizers, except for the mineral fertilizer (Table 4.12). That was probably related to an increase in the soil cation exchange capacity favoured by the biochar component (Liang *et al.*, 2006b). The lower potassium content in the soil compared to other applied fertilizers during the second production year could be related to a more efficient uptake of this element by plants during the first year (Haynes and Williams, 1993). Additional research is required on the fate of potassium in the soil after application of the lacto-fermented mix supplemented by biochar.

4.4.7 Soil bulk density

The soil bulk density was positively affected by the application of the lacto-fermented mix supplemented by biochar compared to mineral fertilizer and stored faeces (first year, 0-20 cm), mineral fertilizer and stored urine (second year, 0-20 cm) as well as mineral fertilizer, stored, faeces, urine and cattle manure during the second year (20-40 cm). Research indicates that around 40 % of the crop yield depends on the soil physical conditions (Krupenikov, 1967; Scerbacov and Vassenev, 2000; Berezin and Gudima, 2002). This could be related to an improvement in the soil organic matter and its dilution effect on the soil bulk density, as organic matter is mixed with the denser mineral fraction of the soil. A compilation of different studies by Khaleel *et al.* (1981) demonstrated a linear regression (r = 0.69) between the increase of soil carbon and the percentage reduction of the soil bulk density.

Studies have shown that only slight changes in bulk density can already contribute to better root development and indirect beneficial effects for root penetration, increased soil infiltration capacity and decreased runoff volumes (Mbah and Onweremadu, 2009; Lawal and Girei, 2013). An improvement of the root penetration increases the capacity of the plants to take up water and necessary nutrients for plant growth and development, which in turn stimulates plant biomass and yield (Mbah and Onweremadu, 2009), as might have happened for the applied lacto-fermented mix supplemented with biochar and the lacto-fermented mix alone. The results are in line with other studies on the application of bokashi soil conditioner (a mixture of cattle manure with water, molasses, effective microorganisms and rice bran), which contributed to a reduction in the soil bulk density, and an increase in the soil permeability and porosity as well as nutrient availability (Xiaohou *et al.*, 2008). Increased soil bulk density and compaction, for example as in the case of mineral fertilizer, could have led to a number of negative effects, e.g. decreasing the soil respiration and worsening of the soil borne diseases in the root zone during drier or wetter conditions (Batey and McKenzie, 2006).

4.4.8 Effect of supplementary treatment by vermi-composting

In spite of the beneficial effects of the vermi-composted lacto-fermented mix and biochar on the germination index and root length (e.g. the root length exceeded by 35% that of the control under laboratory conditions, Andreev *et al.*, 2014), the current study did not show the additional treatment via vermi-composting was advantageous compared to the lacto-fermented mix supplemented by biochar, since most of the growth parameters and the corn yield after application of this fertilizer were significantly lower (Table 4.7). It had a positive effect on the soil humus content, but this was not significant as reflected by Table 4.11. However, a single year field application is insufficient for a generalization.

4.5 Conclusions

Two years of application of a lacto-fermented mix of faeces and bio-waste supplemented by biochar on corn growing on a clay-loamy cernoziom showed beneficial effects on corn growth, especially on plant height compared to the control, stored faeces, urine, cattle manure.

During the first year, the nitrate content of plots fertilized by lacto-fermented faeces and bio-waste supplemented by biochar was significantly higher than that of a control and plots fertilized by stored urine, but significantly lower than the applied mineral fertilizer. The higher corn yield and yield components produced might be attributed to the potential of the biochar to prevent nitrate leaching as well to the improvement of soil quality parameters such as organic matter and soil bulk density. Soil conditioners obtained by lacto-fermenting faeces and bio-waste with biochar supplementation might thus be an appropriate strategy for improving soil fertility and crop yield as well as preventing nitrate losses, particularly under unfavourable weather conditions, such as extended droughts followed by rainfall flushes.

The corn fertilized by this soil improver had an increased number of kernels per row and kernels per ear compared to the unfertilized control, stored cattle manure, faeces and urine, which also improved the yield. Compared to mineral fertilizer, the yield was significantly higher during the first production year, but not significantly different during the second production year. Potassium availability was significantly increased during the first production year in the soil layer 20-40 cm and the soil bulk density was lowered during both production years. In low productive soils and those poor in organic matter content, the lacto-fermented mix supplemented by biochar may serve as an important component in restoring the productivity by improving the soil organic matter and physical soil properties (e.g. lowering bulk density) as well as by preventing nutrient losses. Long term field research is required to fully understand the sustainability of this application, especially on the changes occurring after application of this fertilizer on the soil organic matter and nutrient availability.

Chapter 5. Lactic acid fermentation of human urine for improving its fertilizing value and reducing odour emissions in urine diverting dry toilets

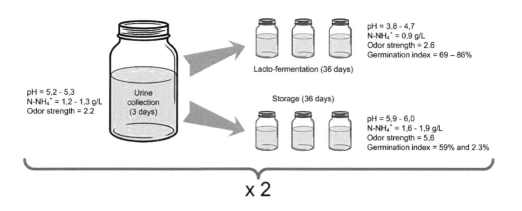

This chapter is based on:

N. Andreev, M. Ronteltap, P.N.L. Lens, B. Boincean, M. Wernli, E.Zubcov, N. Bagrin, N. Borodin Lactic acid fermentation of human urine for improving its fertilizing value, Journal of Environmental Management,198, 63-69.

Abstract

During storage of urine, urea is biologically decomposed to ammonia, which can be lost via volatilization and cause malodor. Lactic acid fermentation of urine can decrease nitrogen volatilization and reduce odour emissions. Fresh urine (pH = 5.2 - 5.3 and NH_4^+-N = 1.2 - 1.3 g L^{-1}) was lacto-fermented with bacterial inoculum from sauerkraut and compared to urine stored for 36 days in glass jars. In the lacto-fermented urine, the pH was reduced to 3.8 - 4.5 and the ammonium content by 22 - 30 %, while in the untreated urine the pH rose to 6.1 and its ammonium content increased by 32 %. The concentration of lactic acid bacteria (LAB) in lacto-fermented urine was 7.3 CFU ml^{-1}, suggesting that urine is a suitable medium for LAB growth.

The perceived odour was twice stronger in the untreated urine than in the lacto-fermented one. Lacto-fermented urine induced higher radish germination than stored urine. Adding LAB inoculum to storage tanks in a urine diverting dry toilet reduced the pH from 8.9 to 7.7, while the ammonium content increased by 35 %, probably due to the high initial pH of the urine. As the hydrolyzed (stored) urine has a high buffering capacity for an efficient urine lacto-fermentation, the LAB inoculum should be added to urine storage tanks before urine starts to accumulate there.

5.1 Introduction

During the last decade, a global concern has risen over a more efficient utilization of nitrogen from the application of nitrogen-based fertilizers without producing adverse environmental impacts. Alternative fertilizers capable to replace or complement mineral fertilizers need to be considered. Human urine is recognised as a potentially good fertilizer owing to its high nutrient content and low hygienic risk (Höglund et al., 2002). In fresh human urine, the main part of nitrogen (75-90%) is in the form of urea $[CO(NH_2)_2]$, but also as uric acid and amino acids. Only a small proportion (up to 7 %) is in the form of ammonia (Kirchmann and Pettersson, 1994). Besides nitrogen, urine also contains phosphorous ($H_2PO_4^-$ and HPO_4^{2-}) and potassium (K^+) in ionic forms, calcium (Ca^{2+}), sulphate (SO_4^{2-}) and soluble organic matter (Lind *et al.*, 2001; Maurer *et al.*, 2006), which potentially have a fertilizing value as well.

Urine diverting dry toilets (UDDT) are ideal systems for harvesting urine for fertilizing applications. However, misuse of UDDT systems may cause faecal cross-contamination of the urine. In order to reduce the pathogens to a safe level, urine has to be stored for 1-6 months (WHO, 2006; Jaatinen *et al.*, 2016). During storage, under the influence of bacterial activity and particularly the urease enzyme, urea is degraded to ammonia, which can volatilise (Mobley and Hausinger, 1989; Udert *et al.*, 2006). Urine hydrolysis and urea decomposition are pH dependent. Urease activity is optimal at pH 7-8 (Schneider and Kaltwasser, 1984; Krajewska *et al.*, 2012), while it is inhibited at low pH < 5 (Larson and Kallio, 1954; Schneider and Kaltwasser, 1984). At pH 8.9-9.0, 95% of the nitrogen in the stored urine is in the form of ammoniacal nitrogen (Kirchmann and Pettersson, 1994). As a result of alkalinisation and an increase in the concentration of compounds such as bicarbonate and ammonia, the urine buffering capacity increases as well (Udert *et al.*, 2006).

Besides impacting the efficiency of nitrogen recovery, ammonia volatilization causes undesirable odour emissions in separate collection sanitation systems. The odour is intensified by other malodourous components such as volatile fatty acids, formed under the influence of bacterial activity (Zhang *et al.*, 2013). The odour emissions are annoying for the toilet users as well as for the residential areas in the vicinity where urine is applied onto agricultural fields.

Different methods have been proposed to reduce ammonia volatilization and inhibit urea decomposition in the urine. A successful result of the inhibition of urea decomposition by maintaining a low pH < 4 was obtained by adding strong acetic and sulphuric acids (2.9 g L^{-1}) (Hellström *et al.*, 1999). A limitation of this method is the economical aspect and the health risks during handling of the acids (Maurer *et al.*, 2006). There is also a lack of research on the impacts of acidified urine on soil and crops after its application.

An alternative method for stabilizing nitrogen is biological nitrification with the use of ammonia and nitrite oxidizing bacteria (Udert and Wächter, 2012). However, maintaining bacterial activity in high strength ammonia solutions as urine is challenging and requires skilled handling (Udert and Wächter, 2012). Thus, there is a need to develop a cost effective method to acidify urine that does not affect its fertilizer value and land applicability.

This study, therefore, focused on the efficiency of lactic acid fermentation of source separated urine for increasing its resource oriented potential. The change in pH, ammonium content, buffering capacity, odour reduction, chemical oxygen demand (COD) and potential biological effects on plants of lacto-fermented urine was compared to that of untreated, stored urine. This expands the knowledge regarding such applications in sanitation which is currently limited.

5.2 Material and methods

5.2.1 Experimental set-up

Storage and lactic acid fermentation of fresh urine was performed under laboratory conditions in two trials, each with three replicates, over two periods of 36 days each, during December 2015 - January 2016 and April - May 2016. Urine samples were collected from 2 donors (a female, aged 44 and a child boy, aged 7) for a period of three days and stored in 1 L glass jars tightly closed with a plastic lid. At the end of the collection period, all urine of both donors was thoroughly mixed, chemical analysis was performed and then separated into two parts. The first part was mixed with a lactic acid bacteria (LAB) solution (1:1) and lacto-fermented in glass containers for a period of 36 days. The second part was stored in parallel in tightly closed glass containers for the same period of time as the lacto-fermented urine.

For obtaining the LAB solution, cabbage was fermented over a one month period, after which juice was collected, mixed with sugar beet molasses and water at a proportion of 1:1:9 and kept in a closed plastic jar until the pH dropped to below 4.

After the treatments, chemical analysis was performed in both the lacto-fermented and stored urine samples. In addition, the LAB solution as well as the lacto-fermented urine was analysed for their *E.coli* and lactobacilli concentration. No analysis of *E.coli* was performed in the untreated, simple stored urine.

The efficiency of lactic acid fermentation was also evaluated under real field conditions in a functional household urine diverting dry toilet (UDDT) storage tank in the vicinity of Chisinau (Moldova), used by a family of two adults (male 45 years and female 44 years) and one boy child (7 years). In this test, the 300 L plastic urine storage tank and the urine pipes were thoroughly washed and rinsed with vinegar prior to the experiments. Then, urine was collected in the tank for a period of one week (for pH and ammonia analysis) after which LAB bacterial inoculum and molasses were added in the tank (2.5 kg of molasses was added to 2.5 L of sauerkraut juice and mixed with 10 L of water). This was applied to 130 L of urine accumulated from the 3 people during 36 days. This ratio was considered taking into consideration the optimal ratio obtained from the laboratory experiments capable to reduce the pH below 4. During each toilet use, the urinal and urine compartment were sprinkled with a LAB solution, which was approximately 70-100 ml per day.

5.2.2 Odour evaluation

The odour intensity of lacto-fermented and stored urine was evaluated by a panel of four people (2 males and 2 females) independent from each other. Perception of the strength of the perceived odour was evaluated according to a rank scale from 0 (no odour) to 6 (extremely strong odour) as described in Table 5.1 (Misselbrook *et al.*, 1993).

Table 5.1 Rank scale for different perceived urine odour strengths

Perceived odour strength	Rank scale
No odour	0
Very faint odour	1
Faint odour	2
Distinct odour	3
Strong odour	4
Very strong odour	5
Extremely strong odour	6

5.2.3 Germination tests

The treated urine samples were diluted 1:10 with distilled water. Twenty seeds of radish *Raphanus sativus* were placed on Petri dishes lined with Whatman filter pads and an amount of 3 ml of 1:10 urine:water mix was added to each replicate. As control 3 ml of distilled water was used. The urine dilution rate (1:10) was obtained after preliminary germination testing and was required because of the low pH of the lacto-fermented urine. After 72 hours, germination was stopped by adding 3 ml of 50% alcohol to each of the Petri dishes. The germination index (GI) was calculated according to the following formula (Eq.1):

$$GI = (G * RRG) * 100 \qquad (1)$$

where G - number of seeds germinated in the sample/number of seeds germinated in the control and RRG - relative root growth (mean root length in the treated sample/mean root length in the control) (Tiqua *et al.*, 1996; Mitelut and Popa, 2011).

5.2.4 Microbiological analysis

The LAB solution and lacto-fermented urine were thoroughly mixed and serially diluted 8 times. Subsequently, 1 ml of the solution was incubated by inoculation on agar plates. *E.coli* incubation was done on HiChrome Coliform agar at 43°C for 24 h, while lactobacilli incubation was done on M.R.S agar (CMO361) at 36 °C for three days under anaerobic conditions (ISO, 1998; MHRF, 2005). The analysis was performed in duplicates.

5.2.5 Chemical analysis

The ammonium (NH_4^+-N) concentration was determined by a UV-VIS Analytik Jena Specord 210 spectrophotometer at 400 nm using cuvettes of 10 mm according to the standard SM-SR-ISO7150-1:2005 (Anon, 2005). Urine was diluted 1000 times (0.5 ml to 500 ml of distilled water) from which 50 ml was taken and added with Seignette salt ($C_4H_4KNaO_6$) and Nessler's reagent.

Chemical oxygen demand (COD) was analysed by dichromate oxidation, using the closed reflux method (Aliokin *et al.*, 1973). Buffer capacity was determined by measuring the initial pH in urine using a Hanna portable EC/pH meter and titrating with 0.1 mol NaOH until the pH changed by one unit (Kirchmann and Pettersson, 1994). All chemical analyses were conducted in triplicates.

5.3 Results

5.3.1 Concentration of Lactobacillus and E.coli in LAB solution and lacto-fermented urine

The microbiological analysis indicated high bacterial counts of *Lactobacillus* both in the LAB solution added to the urine as well as in the lacto-fermented urine (Figure 5.1). The

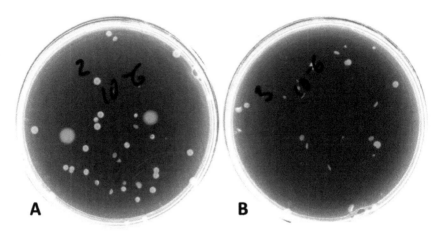

Figure 5.1. Plate counts of LAB (10^{-6} dilution) in the LAB solution (A) and lacto-fermented urine (B).

Lactobacillus concentrations were, respectively, 7.5 and 7.3 log CFU ml^{-1} showing that non-hydrolyzed urine offered favourable growth conditions for LAB. *E.coli* was not detected in the LAB solution nor in the lacto-fermented urine. *E.coli* may appear in urine in the case of urinary tract infection or potential faecal contamination during urine collection (Kunin *et al.*, 1992; Höglund *et al.*, 2002). Under the influence of LAB, *E. coli* growth is inhibited because of the low pH and excretion of inhibitory substances, such as bacteriocins, lactic acid, hydrogen peroxide, glucose oxidase and other compounds (Saranraj, 2014).

5.3.2 Changes in chemical composition of urine by lactic acid fermentation

Lactic acid fermentation did not change the buffer capacity of the urine; however, urine storage increased the buffer capacity approximately two times (Table 5.2). Immediately after the addition of the LAB solution to the urine, the ammonium content was slightly reduced, probably by the dilution (1:1). The pH of the urine during lactic acid fermentation decreased to 4.5-3.8 due to the formation of lactic acid. At this pH value, the bacterial urease is normally inhibited and urea hydrolysation stops (Larson and Kallio, 1954; Schneider and Kaltwasser, 1984). It is important to note that this is only effective with fresh urine; the hydrolysis process is non-reversible, so the lactic acid fermentation needs to take place before ureolysis sets in.

The ammonium content in lacto-fermented urine decreased by approximately 22-30% compared to the fresh urine (Table 5.2). In the untreated urine, the hydrolysis was not inhibited, hence the ammonium content increased by 32% compared to the fresh urine and by 44-53% compared to the lacto-fermented urine. The pH changed only slightly compared to the fresh urine (5.94-6.02 compared to 5.2-5.3, respectively).

Table 5.2 Effect of lactic acid fermentation on pH, buffer capacity and ammonium concentration of urine (mean ± SD).

Sample type	Buffer capacity, mmol L^{-1}	NH_4^+-N g L^{-1} I	NH_4^+-N g L^{-1} II	pH I	pH II	GI I (%)	GI II (%)
Fresh urine	0.9±0.01	1.3±0.05	1.2±0.02	5.3	5.2	-	-
Urine+LAB	0.9±0.01	-	1.1±0.03	4.5	4.4	-	-
Lacto-fermented urine	0.9±0.01	0.9±0.05	0.9±0.1	4.7	3.8	86	74
Stored urine (glass jar)	1.5±0.01	1.9±0.05	1.6±0.1	6.0	5.9	2.2	31.2
Stored urine (urine tank)[1]	-	-	2.9±0.08	-	8.9	-	-
LAF urine (urine tank)[2]	-	-	4.6±0.09	-	7.7	-	0.17
Stored urine (one month, urine tank)			9.6±0.02				

[1] before lactic acid fermentation, [2] after lactic acid fermentation, GI - germination index, I - first experimental run; II - second experimental run.

In the urine tank urea hydrolysation occurred at a much faster rate compared to that in the tightly closed glass containers, since the pH increased rapidly to 8.9 during one week only. The addition of the LAB inoculum and molasses in the urine tank of a UDDT contributed to a reduction of the urine pH by 1.25 units; however, it did not stop the hydrolysation process. Even though the pH was reduced to 7.7 after 36 days of lactic acid fermentation, the ammonium content continued to increase and was 1.5 times higher than the initial value and 3 times higher than in the stored urine (Table 5.2). However, compared to previous one-month stored urine, the amount of ammonia was almost twice lower.

The addition of LAB to urine contributed to a change in the COD concentration (Table 5.3). During lactic acid fermentation, the soluble carbohydrates were converted to lactic acid leading to a pH decline and inhibition of decomposition of organic compounds (Murphy *et al.*, 2007). Therefore, the COD value in lacto-fermented urine was expected to remain similar to that of fresh urine. In contrast, it was 15 % higher (Table 5.3). Also in the untreated, simple stored urine there was no COD reduction, but a slight increase of approximately 7 % compared to the fresh urine (Table 5.3).

Table 5.3 Effect of lactic acid fermentation on COD of urine

Sample	COD, g O_2 L^{-1}
Fresh urine	21.1 ± 0.9
LAB+urine[1]	21.5 ± 1.6
Lacto-fermented urine (glass jar)	25.4 ± 3.0
Stored urine (glass jar)	22.8 ± 0.8

[1] measurement was done immediately after the addition of the LAB solution

5.3.3 The effect of lacto-fermented urine on seed germination

Lacto-fermented urine had beneficial effects on the germination of seeds of *Raphanus sativus* relative to the control, with a GI of 74-86 %. The stored urine with an ammonium content of 1.6 g L^{-1} had a beneficial effect (GI =31.2%) on the germination (Figure 5.2 A). In contrast, the stored urine from the glass jars with a higher ammonium content (1.9 g L^{-1}) had an inhibitory effect on germination and the GI was only 2.3% (Figure 5.2 B). In the urine tank, with an ammonium content of 4.6 g L^{-1}, the germination was also inhibited: the GI was only 0.008 % of that of the control (Table 5.2), probably related to the toxicity of ammonia and ammonium.

Another factor that might have influenced the GI is the pH. Lacto-fermented urine had a higher GI in the first experimental run (pH = 4.7) than in the second run (pH=3.8). The pH of the lacto-fermented urine with a GI of 74% after the dilution of lacto-fermented urine with water (1:10) was increased to only 4, such low pH values are usually unfavourable to germination, while the pH of the stored urine was 6.3 (1st run), which favoured germination.

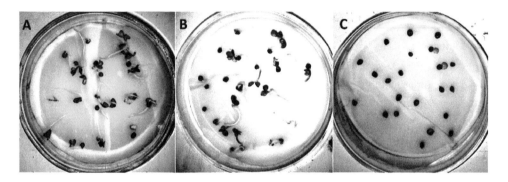

Figure 5.2. Seed germination in control (A), lacto-fermented urine (B) and stored urine (C) (one out of 3 replicates per each sample).

5.3.4 Odour reduction during urine lactic acid fermentation

The odour of fresh urine was perceived as faint to distinct, that of the lacto-fermented urine distinct to faint and that of the stored urine from very strong to extremely strong (Table 6.4). In the toilet room, the urine odour was reduced compared to when no LAB solution was used for rinsing the urine compartments. In addition to this evaluation by different odour panelists, the odour strength in the fresh, lacto-fermented and stored urine could also be perceived during the laboratory analysis. When diluted with distilled water, the fresh and lacto-fermented urine did not produce any nuisance, the odour almost dissapeared after dilution. Instead, the odour of stored urine, particularly in the urine tank, was stronger and smelled even after dilution. Additionally, it was observed that after 36 days of lactic acid fermentation, the pH of lacto-fermented urine had decreased to 3.5 and the urine odour was replaced by a medicinal/ester odour type.

Table 5.4 Strength in the perceived odour of stored and lacto-fermented urine

Sample type	Perceived odour strength	Rank scale
Fresh urine	Faint odour - distinct odour	2.2 ± 0.5
LAF urine	Distinct-faint odour	2.6 ± 0.5
Stored urine (glass jar)	Very strong - extremely strong odour	5.6 ± 0.5
Stored urine (urine tank)	Extremely strong odour	6.0 ± 0.0

5.4 Discussion

5.4.1 Urine as a suitable growth medium for LAB

This study showed that fresh urine can serve as a suitable growth medium for LAB as their number in both the bacterial inoculum and lacto-fermented urine was higher than 7 log CFU after 36 days of incubation. The complex proteolytic system of LAB (Savijoki *et al.*, 2006) probably allowed them to adapt to grow in the urine. Non-hydrolyzed urine is rich in important components such as peptides, urea, hippuric acid, amino acids, citric acid and minerals such as K^+, Na^+, Mg^+, PO_4^{3-}, SO_4^{2-} and Cl^-, which support or can stimulate the growth of LAB (MacLeod and Snell, 1947; Strong *et al.*, 2005; Udert *et al.*, 2006). Other components in non-hydrolyzed urine detected in small quantities in healthy people are carbohydrates such as glucose, lactose, galactose, lactose, xylose and arabinose (White and Hess, 1956; Date, 1958).

As LAB are not able to synthesize all amino acids by themselves, they need additional amino acids and peptides in the growth medium (Niamsiru and Batt, 2000). Both non-hydrolyzed urine and molasses, added as a carbohydrate source for LAB, contain most of the essential amino acids required for the growth of the LAB and the production of lactic acid

(Dunn *et al.*, 1947; Stein and Carey, 1953; Mee *et al.*, 1979) (Table 5.5). Amino acids and peptides can also serve as a nitrogen source, since LAB cannot catabolise mineral nitrogen (Saeed and Salam, 2013). During the current study, the NH_4^+-N concentration has decreased as a result of lactic acid fermentation, one explanation could be that some free NH_3 was immobilized into the bacterial cell wall or precipitated as ammonium lactate (Ikawa and Snell, 1960; Kuromiya *et al.*, 2010), thus decreasing the NH_4^+- N concentration.

Table 5.5 Amino acids excreted in human urine and present in molasses and those essential for lactic acid bacteria (LAB) growth.

Amino acid	Range value excreted in urine, mg/day	Composition ratio (%) in cane molasses	Essential for LAB growth
Histidine	113 - 320	0.57	+
Methyl histidine	17 - 384	-	-
Taurine	86 - 294	-	-
Glycine	110 - 199	3.33	+
Aspargine	34 - 92	-	-
Serine	27 - 73	3.33	+
Alanine	21 - 71	13.3	+
Threonine	22 - 53	2.19	+
Tyrosine	15 - 49	3.33	+
Lysine	7 - 48	0.21	+
Phenylalanine	9 - 31	0.65	+
Leucine	13 - 28	2.19	-
Isoleucine	13 - 26	1.19	+
Glutamic acid	8 - 40	17.73	+
Cystine	10-21	2.19	+
Arginine	< 10 - 18	1.09	+
Aspartic acid	< 10 - 17	38.76	+
Methionin	< 5 - < 10	0.57	+
Valine	4 - 10	6.62	-

Sources: Duun *et al.*, 1947; Mee *et al.*, 1979; Stein & Carey, 1953.

5.4.2 Benefits of urine lactic acid fermentation

5.4.2.1 Urine lactic acid fermentation and its potential fertilizing value

Ammonia and bicarbonate formed during the process of urea hydrolyzation under the influence of urease positive bacteria contribute to an increase in the buffering capacity (Udert *et al.*, 2006). Due to its high buffering capacity, it is not economical to treat urine with acids to lower NH_3 volatilization (Udert *et al.*, 2006) and thus increase its fertilizing value. A combination of high buffering capacity, increased pH and high NH_3 (NH_4^+-N) concentration might also have a negative impact on soil bacterial nitrification processes. This is caused by inhibition of nitrite oxidation and facilitating its accumulation in the soil (Burns *et al.*, 1995). Nitrite accumulation is undesirable due to its potential phytotoxicity (Beauchamp, 1988). Another negative effect could be the loss of ammonia from the soil following urine application, particularly in soils with high pH and high buffering capacity as well as low cation exchange capacity (Sherlock, 1984). For example, a lower soil cation exchange capacity will allow a smaller percentage of NH_4^+-N cations to bind to the exchange sites.

Urine lactic acid fermentation in this study reduced the buffer capacity, pH and ammonium content. A drop of the pH below 5 usually decreases the urease activity (Larson and Kallio, 1954; Schneider and Kaltwasser, 1984). The formation of free NH_3 and its loss via volatilization is also pH dependent. The highest NH_3 concentration is formed and subject to loss at a pH range 7 to 10 (Hartung and Phillips, 1994) as was obtained in the UDDT urine tank, while at pH 4.5 and below as obtained with urine lactic acid fermentation in the bottles (Table 5.2), no free ammonia is formed (Hartung and Phillips, 1994; Williams *et al.*, 2011). Even though the addition of the LAB inoculum in the urine tank of a UDDT did contribute to a decrease in pH from 8.9 to 7.7, it has not stopped urea hydrolysis and thus the increase of the NH_4^+-N/NH_3 concentration (Table 5.2). However, if comparing to a previous analysis of the one month stored urine, the concentration was almost twice lower. Additional research is required to test if the lactic acid fermentation technique can be effective in preventing urea hydrolysis before any urine starts to accumulate there.

Protein type compounds such as amino acids and peptides from urine and molasses are more efficiently used by the proteolytic system of LAB (Niamsiru and Batt, 2000) than urea (Carvalho *et al.*, 2011). Therefore, nitrogen in urine, after its lactic acid fermentation, will remain available mainly as urea. Urea based fertilizers are among the most frequently used nitrogen fertilizers with more than 50 % of all nitrogen usage at a global level (Glibert *et al.*, 2006), owing to low production cost, high nitrogen content and widespread availability (Hawke and Baldock, 2010). However, concerns are arising over the efficiency of nitrogen in the soil system, due to loss of NH_3 immediately following urea fertilizer application (Hawke and Baldock, 2010). Urea applied to the soil undergoes hydrolysis to NH_4^+-N under the influence of soil urease and is subsequently partly lost as NH_3. Ammonia might also have adverse effects on seed germination (Bremner and Krogmeier, 1989). We assume that applying lacto-fermented urine will not cause increased ammonia volatilization since LAB and their associated pH decrease may act as urease inhibitors in the soil, where urea hydrolysis was inhibited (Table 5.2).

This study showed that the urine with the highest ammonium content completely inhibited the germination (Table 5.2). This is probably related to the toxicity of the increased NH_3 and NH_4^+-N concentrations. The stored urine kept in the glass jars with increased ammonium content (1.9 g L-1) inhibited germination (Table 5.2), thus suggesting a potential threshold value of ammonium phytotoxicity. Uptake of high concentrations of NH_4^+-N by plants may cause intracellular pH disturbance as well as reduction of synthesis of organic acids and plant growth hormones such as cytokinin (Britto and Kronzucker, 2002). In contrast to untreated stored urine, lacto-fermented urine showed more beneficial effects on germination (Figure 5.2 A and B). The action of LAB from lacto-fermented urine and the metabolites they produce may bring additional benefits to the plants and soil. For instance, LAB contributed to a 2-4 fold increase in tomato fresh weight of fruits than the unamended control plants (Hoda *et al.*, 2011) and significantly higher growth of cabbage (Somers *et al.*, 2007). LAB may also exert a range of beneficial effects on soil such as solubilization of the water insoluble phosphate compounds present in the soil, thus increasing their availability to plants (Zlotnikov *et al.*, 2013). Additionally, LAB can suppress soil pathogens, e.g. bacteria and fungi (Visser *et al.*, 1986; Hoda *et al.*, 2011; Murthy *et al.*, 2012) by producing different compounds with antagonistic activity such as organic acids, hydrogen peroxide, cyclic peptides, and phenolic or proteinaceus compounds (Hoda *et al.*, 2011; Fhoula *et al.*, 2013). A better understanding of the mechanisms of urine lactic acid fermentation on nitrogen volatilization, plant uptake and the benefits of LAB on biological, physical and chemical soil components can be achieved in long term field studies that are currently lacking.

An aspect that needs further investigation is the use of lacto-fermented urine for nutrient charging of biochar to form a slow release fertilizer (Schmidt *et al.*, 2015). Studies have shown that ammonia and phosphate absorption onto biochar is more effective at a lower pH, with a maximal phosphate absorption at pH 2.0-4.1, but minimal at pH higher than 6 (Yao *et al.*, 2011; Spokas *et al.*, 2012). Also the adsorption mechamism requires further research. Besides direct adsorption on the carbon structure of biochar, the organic compounds present in urine can form an organic coating to which anions and cations such as phosphorous and ammonium can bind (Schmidt *et al.*, 2015).

5.4.2.2 Urine lactic acid fermentation for odour reduction

Offensive odours in urine formed over time are produced as a result of bacterial metabolism or thermal reactions (Troccaz *et al.*, 2013). Ammonia volatilized during urine storage and urea hydrolysis is a major malodourous compound (Zhang *et al.*, 2013). However, other organic compounds with lower odour threshold values than ammonia play also a significant role in the formation of the typical stale odour of urine after storage (Troccaz *et al.*, 2013). A range of volatile fatty acids are formed as a result of anaerobic degradation of carbohydrates, proteins, peptides and amino acids from urine by (facultative) anaerobes. Volatile fatty acids are among the key factors contributing to odour emissions. Among them, acetic, propionic and butyric acid have the strongest offensive odours (Zygmunt and Bannel, 2008). Other main odour contributors in the stored urine are dimethyldisufide, methyl mercaptan, ethyl mercaptan, thrimethylamine, phenol and indol (Troccaz *et al.*, 2013; Zhang *et al.*, 2013; Liu *et al.*, 2016). Oxidation of methionine to methional and methanethiol leads to the formation of dimethyldisulphide and methyl mercaptan (Wagenstaller and Buettner, 2013; Liu *et al.*, 2016). Dimethyldisulphide was recognised as one of the most offensive odourous compounds emitted from human urine (Liu *et al.*, 2016). Lactic acid fermentation inhibited the activity of bacteria (Wang *et al.*, 2001) like *Enterococcus faecalis*, *Streptococcus agalactiae*, *Morganella morganii* and *Escherichia fergusonii* that generate key odourous compounds such as dimethyldisulphide, thrimethylamine, phenol or indol in stored human urine (Troccaz *et al.*, 2013).

During lactic acid fermentation, LAB can metabolise amino acids via their enzymatic system into flavour compounds such as alcohols, aldehydes, esters and sulphur compounds (Savijoki *et al.*, 2006). Therefore, the reduction of odour emissions during urine lactic acid fermentation was probable caused by the decrease in ammonia emissions as well as by the synthesis of flavour compounds by LAB. For example, LAB can metabolize citrate from urine into diacetil, acetoin and butanediol, which are important flavour compounds (Hassan *et al.*, 2012). Also, hippuric acid, present in urine, is decomposed completely during lactic acid fermentation, this being the precursor for the synthesis of benzoic acid, which is another flavour compound (Güzel-Seydim *et al.*, 2000).

In contrast to the glass containers, where odour was considerable reduced after lactic acid fermentation, the odour was stronger in the UDDT urine tank (Table 5.4). This was probable caused by the fact that free urease and urease positive bacteria prevailed in the pipes and urine tank. Their complete elimination is impossible, therefore urea hydrolysis and anaerobic decomposition of organic matter did take place and the LAB could not dominate the processes in the urine tank, even though they contributed to a pH reduction (Table 5.2). Another factor that might have hindered urine lactic acid fermentation was that the urine tank was not completely anaerobic, thus other microorganisms besides LAB could compete for organic matter and nutrients from the urine.

This study showed that COD values do not reflect the decomposition of organic compounds from urine (Table 5.3). The increasing, instead of decreasing, COD concentration in the stored and lacto-fermented urine was probably due to the nitrogen organic compound urea, which has no COD value on a molar basis, thus the results being underestimated and explaining the increased value in stored urine where anaerobic bioconversion into organic compounds occurred. The increased COD concentration in lacto-fermented urine might also have been caused by lactic acid, which is not completely oxidized during bichromate oxidation in the COD analysis (Elizarova, 2000). An additional explanation might be that hydrogen peroxide generated by LAB (Kang *et al.*, 2005) consumes the oxidation agent potassium dichromate, leading to an overestimation of the COD concentration. For example, hydrogen peroxide present in anaerobically digested livestock wastewater led to a 9 - 14 % overestimation of the theoretical COD values (Lee *et al.*, 2011).

5.5 Conclusions

Urine lactic acid fermentation might be an effective low-cost technique that may lower ammonia volatilization and reduce odour emissions from UDDT systems. The application of a LAB solution of sauerkraut containing molasses and water to fresh urine led to effective acidification to pH < 4 - 4.5 and a reduction by 1/3 of the ammonium content, while maintaining a high concentration of viable LAB (7.3 CFU ml^{-1}) compared to the stored urine. In addition, lacto-fermented urine reduced by two times the perceived odour strength and beneficially affected seed germination, showing a potentially higher fertilizing value than untreated, stored urine. Applying this technique to an urine tank of a UDDT reduced the pH, however did not stop urea hydrolyzis. Compared to previous studies on the concentration of NH_4^+-N in one month stored urine the ammonium content was twice lower after urine lactic acid fermentation, thus suggesting that ammonia reduction might be possible. However, additional research is required to transfer these research findings to practical application in UDDT systems.

Chapter 6. General Discussion and Outlook

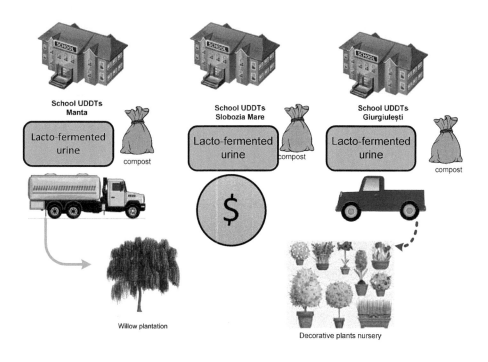

6.1 Potential and limitations of the application of lactic acid fermentation and biochar in excreta treatment

During the last few decades, the growing threats such as depletion of soil nitrogen, phosphorus and organic matter, eutrophication of rivers, competition for land between farmers and other interests (Blum, 2013), loss of soil fertility due to erosion and overexploitation and the challenge of establishing sustainable sanitation call for alternative approaches in sanitation and agriculture. The productive use of excreta from UDDTs can bring many economic and social benefits and contribute to an improvement in the soil quality by substituting or complementing chemical fertilizers. The recycling of nutrients and organic matter from excreta to the soil is becoming extremely important, given the current rate of loss in soil organic carbon stocks (FAO, 2015), combined with the depletion of non-renewable sources of nitrogen and phosphorus fertilizers (Dawson and Hilton, 2011) and the uneven distribution of potassium resources in nature (Ciari *et al.*, 2015). The application of only chemical fertilizers to the soil has proven to be unsustainable, especially for countries that face financial limitations and ever increasing prices of the fertilizers. Long term experiments (Boincean *et al.*, 2014) have shown that the application of increased amounts of chemical fertilizers leads to rapid decomposition of humus and even decreased yields (Mulvaney *et al.*, 2009). Supplementation with organic matter is important for maintaining soil fertility and increasing soil resistance to erosion (Lal, 2002).

Access to safe water and sanitation is a great challenge for the rural and peri-urban areas of Eastern European regions. A significant part of the population lives in rural areas and the connection to sewerage is either unfeasible or unaffordable in the immediate future. The abundant existing pit latrines in these regions affect the quality of groundwater sources and the environment (Bodik, 2007). Therefore, dry sanitation, such as urine-diverting dry toilets (UDDTs), can be a sustainable alternative to latrines, as they provide valuable fertilizer and reduce groundwater pollution. Also, in order to avert the water crisis in the future, it is crucial to adopt management techniques for more efficient use and protection of water resources that are applicable under economic limitations as well as weather or hydrological extremes (e.g. droughts).

In Moldova, a significant proportion of the rural population depends on shallow wells for their drinking water, and these get polluted by onsite sanitation systems (pit latrines and septic tanks) and the disposal of animal manure (Andreev and Andreev, 2010). Even though there is connection to sewerage systems in some rural areas, the treatment facilities are either dysfunctional or not existent (The World Bank, 2014), with the wastewater infiltrating into the soil or discharged into the surface waters without treatment. The predictions from the impact of climate change indicate that water supply will be threatened due to scarcity in surface and groundwater resources. Currently a significant proportion of the rural population is already confronted with the issues of insufficient water availability during extended droughts (OECD, 2013).

UDDTs are quite well promoted at the national level (ApaSan, n.d.), are supported by the government (e.g. National Centre of Public Health, Ministry of Environment) and are socially accepted at the local level. The Swiss Development and Cooperation Agency' Programme ApaSan has so far built 52 school UDDTs and more are planned in the future. Also, several pilots of such facilities were built at the household level. UDDTs need further improvement for reducing the required long-term storage of 1-2 years for faeces and 6 months for urine (WHO, 2006) in order to recycle more nutrients and organic matter to the soil and control the odour from the urine compartments.

Soils in the Republic of Moldova are an important economic asset. Maintenance of its productive capacity over the long term is a prerequisite for sustainable development of the country. Once having rich soil resources, Moldova is currently faced with serious dehumification and erosion of soil resources, which greatly affect the economy. The main soils of Moldova (in Russian: chernozioms) are naturally resilient, thanks to their unique physical and mineralogical properties, the thickness of the humus in the topsoil, and their restoration capacity. Nevertheless, having been intensively exploited in the past, these soils have become susceptible to the loss of humus and nutrients, through erosion and compaction.

During the last 30 years, an acceleration of soil erosion and continuous loss of soil organic matter and nutrients has occurred in Moldova (Krupenikov *et al.*, 2011c). In contrast to the past overuse of chemicals, currently there is a limited application of chemical and organic fertilizers, such as cow manure particularly in subsistence farming systems. The use of fertilizers in Moldova has declined substantially compared to the 1980s due to land privatization (Spoor and Izman, 2009), when land was distributed from centrally controlled big collective farms to small plots for farmers without a transfer of knowledge on soil management practices. Twenty times less mineral fertilizer is currently being applied compared to 30 years ago and the application of farmyard manure has declined to only a few hundred kg ha^{-1} (National Bureau of Statistics of the Republic of Moldova, 2016 a). Due to a lack of fertilization, particularly of replenishment of soil organic matter, areas affected by erosion and dehumification continue to expand (Leah, 2012; Andries *et al.*, 2014).

Soil fertility can be restored by enhancing the factors that form soil. One of the important factors is the enrichment of soils with organic matter, which helps to improve the soil quality and to increase the yields. Soil conditioners obtained from a lacto-fermented mix of faeces and bio-waste, supplemented with biochar and lacto-fermented urine would be inexpensive alternatives to chemical fertilizers. Co-composting faeces with waste from the food and beverage industry is an interesting option worth exploring. For example, waste from the food and beverage industry accounted for approximately 8% of the industrial waste. The amount of waste generated during 2015 was 309,000 tons, out of which only 19% is used; 16 % of this waste was landfilled and the rest mainly remained on the properties of these businesses (National Bureau of Statistics of the Republic of Moldova, 2016 b).

This thesis highlights the possibilities and limitations of the application of lactic acid fermentation in excreta treatment (Chapter 2), including the effect of combined lactic acid fermentation and thermophilic composting on pathogen removal (Chapter 3), the effects of co-treated faeces, bio-waste and biochar on corn growth, yield, yield components and soil quality (Chapter 4), and the treatment of urine via lactic acid fermentation for ammonia reduction and odour control (Chapter 5). Based on the results of the current research, an overview is made of the application of lactic acid fermentation combined with composting (thermophilic composting or vermi-composting) and biochar addition for the treatment of human excreta to improve resource-oriented sanitation using UDDT and its potential reuse in agriculture or horticulture. The main advantages and limitations of lactic acid fermentation of human excreta are explained in Chapter 2, which is a useful guide to help practitioners to better understand the use of this microbiological process for improving the functionality of UDDTs, together with better nutrient recycling.

The findings reported in Chapter 3 outline the safe treatment of human faeces through the use of lactic acid fermentation combined with thermophilic composting for pathogen removal. The effect of this compost on the germination and growth of plants is also explored in this chapter to understand the potential for the recycling of the nutrients present in the human faeces that are collected in UDDTs. The pathogen removal in lacto-fermented excreta (Table 3.3) did not comply with the recommendations for soil amendments that are obtained from biological treatment (Hogg *et al.*, 2002), aside from improved hygienization through supplementary treatment via thermophilic composting (Andreev *et al.*, 2017).

An interesting fact is that the sanitizing temperature of 56-65°C during the post-treatment composting stage, after the lactic acid fermentation, was achieved without turning or mixing (Andreev *et al.*, 2017). This may be related to the types of microorganisms that are involved in this process under low oxygen conditions (Wang *et al.*, 2007b), the enzymes synthesized by lactic acid bacteria during fermentation (Muller, 2001), or the concentration of easily degradable organic carbon in the substrate (Hayes and Randle, 1968). This research suggests that the treatment of excreta via lactic acid fermentation is best carried out at the post-collection stage, in combination with other types of waste such as cattle manure or fruit waste. This will also help prevent other environmental problems since food waste can easily get spoiled, generating offensive odour (Qu *et al.*, 2015), contributing to gas emissions (Qu *et al.*, 2015) and attracting disease vectors (e.g. rodents or insects) (Drechsel *et al.*, 2015). For example, according to EPA estimations, during 2012 only 3 % of food waste was composted in the USA, while the rest was landfilled, generating approximately 18 % of all US methane emissions (Tim, 2015).

Chapters 4 provides valuable insight on potential agricultural applications of lacto-fermented faeces and bio-waste, supplemented by urine charged biochar. This research highlights the effects of faecal based soil conditioner on corn growth, yield, yield components and soil quality (Chapter 4, Tables 4.4-4.10). An emphasis is made on the role of such type of soil conditioners under the conditions of decreased soil organic matter and increased drought frequency. This research revealed that corn fertilized with lacto-fermented faeces, bio-waste and biochar soaked in urine had a significantly higher yield ($p < 0.05$) than an unfertilized control and plots fertilized by stored urine, faeces, cattle manure, and even chemical fertilizer (during the first production year).

Higher nitrogen availability, coupled with the capacity of biochar to retain water, had probably allowed for better root development of the corn and more efficient nitrogen uptake during the period of intensive growth, thus making the corn more resilient to the following drought during yield formation (Spiertz and De Vos, 1983). This aspect could be clarified further with additional research on nitrate uptake by corn during the period of intensive growth (June-July) and the translocation of nitrate from vegetative parts of the plant to the grain, during the grain formation (August-September) following the application of lacto-fermented faeces, bio-waste and biochar, and the chemical fertilizer (Chapter 4). These data also need to be linked to soil nitrate content, soil available water storage capacity, precipitation data (Vronskih, 2014), and estimations on nitrogen loss by leaching (Burgos *et al.*, 2006). The applied lactic acid fermented excreta and biochar has added to an improved organic carbon content in the soil even though the increase was not significant ($p>0.05$, Table 4.13). The experimental length was too short to observe changes in the soil organic carbon concentration. Such changes can only be observed by using C-labeled incubation experiments (Perelo and Munch, 2005). Further, it is also required to explore the role of biochar in the process of lacto-fermentation and composting and its role in nutrient retention. It is important to study further how biochar influences ammonia retention from urine and organic carbon stability in the compost (Glaser, 2015).

Chapter 5 explores lactic acid fermentation of urine. Data on this approach is very limited and is important as it offers useful guidelines for practitioners, particularly those from countries with scarce financial resources, on cost effective techniques for improving the fertilizing value of urine and reducing odour emissions. This study's laboratory research on urine lactic acid fermentation brings new perspectives for reducing the ammonia emissions and odour control in UDDTs. Although further research is needed to further support these findings and to confirm the efficiency in real-life conditions, the available evidence indicates that lactic acid bacteria from common home-made products such as sauerkraut can survive in such a high strength solution as urine, contributing to pH reduction to less than 4, an ammonium concentration decrease and odour reduction (Chapter 5).

In this chapter, recommendations are given for the use of lactic acid fermentation combined with composting and biochar in UDDTs for the treatment of both the urine and faeces fractions for their reuse. Furthermore, practical recommendations on the application of the treated urine and faeces for irrigating willow trees or growing ornamental plants are also given.

6.2 Product quality assurance for potential full-scale application

When there is a demand for large scale production and marketing of a product, appropriate standards which stipulate the conditions for adequate pathogen reduction should be applied (Hogg *et al.*, 2002). In Europe, the application, testing and product quality assurance of organic fertilizers, including compost, are based on statutory and voluntary standards. The statutory standards cover most important characteristics for the protection of the environment and human health (Hogg et al., 2002). Voluntary systems of quality assurance support these standards by making recommendations regarding suitability of products for different enduses. These standards refer to a number of requirements for feedstock materials, the characteristics of the process, end-product quality and protection of soil quality (Figure 6.1). Feedstock requirements categorize the input materials that are allowed in this specific biological treatment. According to the existing regulations at the European level, the input materials like bio-waste, sewage sludge, animal manure, yard waste and agro-industrial by-products are included in the list of acceptable substrates for biological treatment. Source-separated excreta are not listed as feedstocks for biological treatment and there are currently no regulations at the European level (Hogg, 2002).

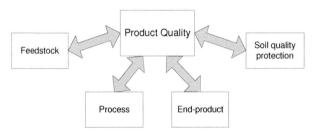

Figure 6.1 Main components to be considered in the development of a standard lacto-fermented fecal compost (Hogg *et al.*, 2002)

Biochar is not included among the recommended feedstocks, but its quality shall be assured according to the guidelines of the European Biochar Certificate (EBC, 2012). The main quality criteria for biochar specified by the European Biochar Certificate versus those investigated in this thesis are reflected in Table 6.1.

Table 6.1 Main quality parameters of biochar as specified by the European Biochar Certificate (Source: ECB, 2012).

Quality Parameter	Required as specified by EBC	Results on wood charcoal used in the current study
C content	> 50 %	60 %
Nutrient content	N, P, K, Mg, Ca	Not investigated
Heavy metals	Pb, Cd, Cu, N, Hg, Zn, Cr, As	Not investigated
Volatile organic compounds, polychlorinated biphenyls	To be specified, threshold values	Not investigated
Other parameters	Bulk density, electric conductivity, pH, ash content, volatile matter, water holding capacity	pH, moisture, ash content, specific surface area, pore volume

It refers to the carbon and nutrient content, the concentration of polychlorinated biphenyls (PCB), volatile organic compounds (VOC), heavy metals and the water holding capacity of biochar. The process specifications refer to the maximum acceptable limits for pathogens, weed and plant propagules. These requirements vary across different countries. A summary of these requirements for pathogen indicators, weed germination and plant tolerance for compost maturity are indicated in Table 6.2.

Table 6.2 Standards for pathogens, weed seeds and propagules in compost.

Indicator	Requirements specified by existing standards[a]	Results from this research
Salmonella	Absent	Absent (Andreev *et al.*, unpublished data)
Parasites (*Ascaris*)	0 in 100 g sample	0 in 100 g sample
Enterobacteriacea	$< 10^3$ CFU g^{-1}	LAF: *E.coli* - < 3 CFU g^{-1}, *Enterococcus faecalis* $< 10^3$CFU g^{-1}
Enterococcus faecalis	$<10^3$ CFU g^{-1}	*E.coli* $< 10^3$CFU g^{-1} after LAF+TC[b] and after LAF+VC[c]
Weed propagule test	5 viable plants L^{-1}	Not investigated in this study
Plant tolerance	20 % below control	LAF - 1-3% above the control, LAF+TC - 10 % below the control, 6 LAF+VC - 16 % below the control

[a] - Hogg *et al.*, 2002
[b] - LAF+TC - lactic acid fermented mix of faeces and bio-waste followed by thermophilic composting.
[c] - LAF+VC - lactic acid fermented mix of faeces and bio-waste followed by vermi-composting.

Even though the lacto-fermented material was not sufficiently sanitized, it had stimulatory effects on the growth of plants. This may be related to the substances synthesized by lactic acid bacteria together with the effect of biochar onto which nutrients from urine were adsorbed. The required time for curing and maturation is not specified in most compost standards. The only exception is the Irish regulation, which stipulates that compost must be cured for at least 21 days (Hogg *et al.*, 2002). In this study, the curing phase of lactic acid fermentation followed by thermophilic composting lasted for only 11 days (Chapter 3, Figure 3.1), which was not in accordance with the recommendations for a mature compost. In addition, both germination tests were within the recommended limits and the tomato growth exceeded that of the control (Chapter 3, Figure 3.2). Nonetheless, additional research on compost maturity indicators, e.g. C:N ratio, respiration level, concentration of stable organic carbon is required to enable safe use of human waste derived compost.

Depending on the type of use (agriculture, horticulture or other), standards on the quality of the final compost specify the nutrient concentrations of N, P and K and the maximum acceptable values for toxic substances and pathogens. Table 6.3 compares the nutrient content of compost treated by aerobic composting, digestate treated by anaerobic digestion (Hogg *et al.*, 2002) and lacto-fermented mix of faeces, bio-waste and biochar (this thesis). Both the nutrient content and the effect on plants growth and yield from radish germination tests (Andreev *et al.*, 2014) and two year field experiments on corn (Andreev *et al.*, 2016) indicated beneficial effects. However, the level of pathogens in the lacto-fermented mix was not in accordance with the requirements of a compost (Table 6.2). In this case, it would be appropriate that application of lacto-fermented faeces, bio-waste and biochar is used for non-food (e.g. horticultural and ornamental) plant cultivation or additional treatment via thermophilic composting is performed. The concentration of heavy metals such as Cd, Cr, Cu, Hg, Ni, Pb, Zn, As, Co is required to be analysed to assess if the compost complies with the certification.

Table 6.3 Nutrient content in compost obtained via aerobic composting, anaerobic digestion and lactic acid fermentation with the addition of wood charcoal loaded with nutrients from urine.

Nutrient	Aerobic compost, kg/tonne^{-1}	Digestate, kg tonne^{-1}	LAF+C[1], kg tonne^{-1}
N	2-4	4-4.5	18.7
P	1-2	0.5-1.0	2.4
K	1-2	2.5-3	Not investigated

[1] - without the content in charcoal

The soil quality determines the application rates of heavy metal and nutrients that can be added the soil (mg kg^{-1} d.m.) with a compost, e.g. what are the admissible loads of heavy metals (kg ha^{-1} year^{-1}) and admissible loads for nutrients, such as nitrogen and phosphorous. This thesis contributed with data on the nutrient content in the soil following application of lacto-fermented excreta and biochar. Based on these data and additional research, it will be possible to calculate the acceptable loading rates of lacto-fermented excreta and biochar for avoiding nutrient pollution, while ensuring a beneficial agricultural effect.

6.3 Input materials required for lactic acid fermentation of excreta

The main input materials (Figure 6.2) included in the lactic acid fermentation of excreta are waste materials available in areas where UDDTs are located. The main limiting input materials are the carbohydrate sources and the LAB inoculum. As lactic acid fermentation is widely used today, with plenty of homemade fermented products (e.g. diary products and vegetables), these can serve as sources of LAB to replace commercial bacterial inocula. Fermented foods are preferred by consumers because of their taste, texture, nutritional values, antitoxin and antimicrobial effects (Mokoena *et al.*, 2016). Application of lactic acid fermentation in UDDTs at household level may increase the need for more lacto-fermented products, thus contributing to a healthy, probiotic nutrition.

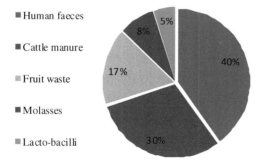

Figure 6.2 Input materials added for lactic acid fermentation of faeces. Biochar soaked in urine (10 % wet weight) was added after 8 weeks of lactic acid fermentation.

One of the limiting factors for scale application of lactic acid fermentation in UDDTs is the need for the addition of sources of soluble carbohydrates or sugar containing substrates, as sugar is the main growth substrate for LAB (Yemaneh *et al.*, 2013). For an efficient lactic-acid fermentation, sources of soluble carbohydrates need to be used that have other competitive economic uses. For example, experiments on excreta lactic acid fermentation were carried out with the application of wheat bran and mollases (Yemaneh *et al.*, 2013; Böttger *et al.*, 2014; Scheinemann *et al.*, 2015; Andreev *et al.*, 2016; Andreev *et al.*, 2017). Low cost substitutes for the input materials have to be sought. Press mud, a sugar byproduct (Figure 6.3) that contains 5-15% of sugar (Xavier and Lonsane, 1994) can be a potential alternative.

Figure 6.3 Press mud - a by-product of the sugar industry and potential cost-effective carbohydrate source for faeces lactic acid fermentation (scale: cm).

It has been shown that this by-product of the sugar industry can be an inexpensive substrate for lactic acid production (Xavier and Lonsane, 1994). Its potential needs to be studied for the use as an additive during the faeces collection stage in UDDTs or at the of lactic acid fermentation stage. In Moldova, this waste product from the sugar industry is piled on potentially agricultural land, thus its availability is not a limitation. Some waste products such as e.g. fruit waste, grass clippings and kitchen scraps can be used as sources of water soluble carbohydrates. Every day, a huge amount of fruit and vegetable waste is generated at the household level, vegetable markets or fresh fruit boutiques as well as by the food processing industry. Food waste can be a substitute for molasses; however, it requires pre-processing, i.e. storage in an oxygen limiting environment to release sugars (Wang *et al.*, 2001). For example, a study carried out at the European level estimated that food waste accounts for approximately 20 to 30% of the produced food (Stenmarck *et al.*, 2016), many of such wastes could be useful for lactic acid fermentation. It would also be interesting to test the application of weed plants which contain elevated concentrations of glucose, fructose, succrose or inulin, e.g. *Cicorium intybus, Sonchus oleraceus, Lactuca seriola, Inula helenium, Helianthus tuberosus* (Van Laere and Ende, 2002; Petkova and Denev, 2012).

The availability of biochar can also be a limiting factor in scaling of excreta lactic acid fermentation, when it is not produced at the local level. One of the options is to develop portable, economically viable small scale biochar kilns (Nsamba *et al.*, 2015) for the farmers who practice integrated agricultural activities with agroforestry to produce their own biochar. For example, a study carried out in Thailand showed that a farmer who grows maize, rice and mixed agroforestry can produce up to 4.68 t of biochar from both field residues and agroforestry prunings (Frogner, personal communication). Biochar selling to interested compost producers can be an additional source of income to the farmers.

Another option can be the wood charcoal produced for griling, if this is not treated with any additives to maintain ignition. For example, in Moldova there are few local enterprises who produce wood charcoal for grilling in self-made metallic pyrolysers (Figure 6.4).

Figure 6.4 Metallic pyrolysers for wood charcoal production in Rezeni (Moldova)

These entreprises sell the charcoal, however the small pieces and dust, which remain can be piled and used for agriculture. The quality of wood charcoal is in acceptable limits that do not harm plants (Andreev *et al.*, 2016). However, it is important to investigate other quality parameters stipulated by the European Biochar Certificate to assess the applicability of this material in agriculture.

6.4 Agricultural application of lacto-fermented excreta and biochar

This research indicates that if even the agricultural effects of lacto-fermented faeces, bio-waste and biochar are positive for both soil quality and hygiene, and organic carbon stability, it is important that the lactic acid fermentation stage is combined with thermophilic composting or biochar is added to the lacto-fermented mix alone in horticulture. Additional research is needed to clarify the level of decomposition and stabilization of organic matter and the degree of nutrient loss at both the lactic acid fermentation and the thermophilic composting stage and how biochar can improve the stabilization of carbon and prevent nutrient loss. For example, research has shown that the labile carbon diffuses into the biochar pores and is adsorbed onto its surface, thus preventing its abiotic and biotic decomposition (Zimmerman *et al.*, 2011). The fertilizing effect of the combination of biochar with lacto-fermented faeces and urine still needs to be investigated in long term field experiments, especially with respect to the mechanisms of nutrient absorption from urine, nutrient availability and nutrient uptake by the crops.

In separate collection systems, urine collection is also of particular importance, owing to the abundance of nutrients and their availability to plants and its larger volume as compared to faeces. The research carried out on the effect of lacto-fermented urine on plants (Chapter 5) is yet at the incipient stage with only germination tests so far and should be supplemented with field experiments. It is important to assess the potential of biochar to

142

absorb nutrients from acidified urine, considering the potentially more efficient nutrient absorption onto biochar at a lower pH (Yao *et al.*, 2011; Spokas *et al.*, 2012).

The lacto-fermented faeces, bio-waste and biochar was found to be a superior fertilizer compared to stored urine, faeces and cattle manure, even though further investigation is required if the latter does improve the soil moisture and humus content. The fate of the mobile phosphorus and potassium in the soil and plants, in addition to the level of nitrogen mineralization in the soil after application of lacto-fermented faces, bio-waste and biochar. The effect of chemical fertilizer differed between the two production years, with no increase in yield during the first year and a significant one during the second year (Figure 4.5, Table 4.8), compared to the unfertilized control. This was probably related to the effect of heavy rainfall on nitrate leaching during the first year (Chapter 4, Figure 4.5, Table 4.8). Research on nitrate leaching and plant uptake after application of lacto-fermented excreta with biochar compared to mineral fertilizers under different rainfall conditions can help to reveal if this soil conditioner can be a fertilizer substitute under different climatic conditions.

6.5 Application of lactic acid fermentation combined with composting and biochar in urine diverting dry toilets in Moldova

6.5.1 Site description

Moldova has a rich experience in implementation of ecological sanitation. During the last 8 years, several pilot household and school urine diverting dry toilets (UDDTs) were implemented by external donors and NGOs. UDDTs were recommended as appropriate decentralized sanitation solutions for rural areas in the Country Document on Target Settings for the Protocol on Water and Health (UNECE, 2011). Moreover, recently, the international association Women in Europe for a Common Future (WECF) in cooperation with Women in Sustainable Development of Moldova (WiSDOM) and local stakeholders, developed practical guidelines on the construction of urine-diverting dry toilet facilities and guidelines for the use of corresponding fertilizers are being prepared. Therefore, suitable conditions exist for real-world testing, setting an example of ecological sanitation for similar countries, including the resource required and technical issues for correctly implementing and operating of UDDTs.

However, many schools where UDDTs have been introduced are facing challenges, particularly regarding the handling of urine and faeces as well as overcoming the odour emissions. Such an example is the school Slobozia Mare, located in the Cahul district, South of Moldova. Through an interview with the school director and the toilet care taker in this school, it was seen that the administration of the school faces challenges, in particular the rapid filling of the urine tank and the lack of experience regarding tank empting and the use of urine as a fertilizer. During the visit, the door of the toilet was closed with a sign saying "entering is strictly prohibited" (Figure 6.5 A), the pupils continuing to use the old pit latrine outside of the school (Figure 6.5 B).

Figure 6.5 A) The closed door of the UDDT in Slobozia Mare, (Cahul, Moldova) and B) Pupils use again the pit latrine in the school yard.

6.5.2 Lactic acid fermentation of faeces followed by thermophilic composting and biochar addition as post-treatment

At the collection stage of the faeces, cover material rich in carbohydrates is added, including a sugar beet factory waste called press mud, chopped corn stalk and grass. This helps to control the odour and promotes the growth of lactic acid bacteria already available in the faeces. In order to reduce the pathogens and stabilize the faecal material, a secondary treatment of faeces is carried out, in which the faeces are mixed with other kinds of bio-waste (kitchen scraps, green refuse or animal manure), molasses and LAB for lacto-fermenting it for a period of 10 days (Chapter 2 and 3). Kitchen and garden waste should be pre-fermented in closed barrels before being mixed with faeces. Next, thermophilic micro-aerobic composting has to be carried out for a period of approximately one month (Chapter 3). The compost becomes mature when the temperature drops to its initial level. Biochar is added at a concentration of 10-15% of the compost weight when the temperature raises above 50 °C to lower the ammonia concentration. When temperature falls below 30-40 °C, at the maturation stage, the compost becomes covered with a white mold (Figure 6.6).

Figure 6.6 The compost obtained via combined lactic acid fermentation and thermophilic composting of faeces and other bio-waste together with biochar, at a maturation phase of 30-40°C, which is characterized by the growth of white mold.

6.5.3 Lactic acid fermentation of urine

Acidification of urine is important for the reduction of ammonia volatilization and odour emissions, effective hygienization and improving the overall agricultural value of urine. The main suggested management steps applied in UDDTs are:

Acidification of urine in the storage tank. For succesful lactic acid fermentation in the urine tank, a LAB solution has to be added before the urine starts to accumulate there, to avoid abundance of urease positive bacteria and to ensure a cost effective lactic acid fermentation. The urine tank should be closed to limit oxygen penetration since lactic acid bacteria are facultative anaerobes. For reasons of hygiene, urine needs to be preserved for 1-2 months before its agricultural application. For the application of the LAB solution in those emptied urine tanks, it is important that the pipes and urine tank are well rinsed with acetic acid or a LAB solution. For example, sauerkraut juice that has been matured with sugar rich liquids, e.g. sugar molasses or whey, depending on the availability. Another cheap alternative for schools is the water remaining after the washing the rice or pots that contained dairy products. This liquid may also be obtained by simply rinsing the spent milk packages. The liquid obtained after rinsing the pipes and urine tanks is not toxic and can be poured on a compost pile or used to water trees.

Acidification of urine within the toilet. After each UDD toilet use, the urinals and urine section of the toilet seat should be sprayed with a LAB solution. Preferably, the LAB should be activated before addition to the urine tank by mixing it with molasses and leaving it for bacteria to grow in this solution for at least one week. The LAB solution shall be preserved in a cool place and not stored longer than 1-3 months.

Potential post-treatment of urine after lactic acid fermentation. Acidified urine can be used to soak on biochar, the latter can be mixed with compost. Alternatively, the acidified urine can be diluted with water (at least 1:10, as suggested by the germination test with lacto-fermented urine, see Chapter 5) and applied to the soil around the cultivated plants.

The costs of acidification of urine by lactic acid fermentation were three times lower than chemical treatment by concentrated sulfuric acid (Hellström *et al.*, 1999). As reflected in Table 6.4, the overall costs of urine acidification via lactic acid fermentation can be at least 3 times cheaper than chemical acidification with sulfuric acid. Note that the cost for the LAB solution includes sauerkraut, salt and molasses and the price of H_2SO_4 is the price of the concentrated acid used in laboratory analysis, not of a highly purified quality. Aside from the economic aspects of the treatment, potential risks and limitations have to be taken into account. While for lactic acid fermentation, the availability of molasses could be the main limitation, the negative side effects in the application of sulfuric acid include the health hazard of handling and transporting the acid, increased H_2S emissions that may negatively impact the soil by raising its electric conductivity (Frost *et al.*, 1990) and increased concentrations of malodourous volatile fatty acids and volatile sulphurous compounds (Ottosen *et al.*, 2009; Petersen *et al.*, 2012).

Table 6.4 Costs and potential risks/limitations for urine acidification by lactic acid fermentation compared to the chemical acidification with concentrated sulfuric acid.

Process	Quantity of additives* required/m³ of urine	Cost/m³ of urine, €	Potential risk or limitation
Urine acidification by lactic acid fermentation	20 L LAB inoculum; 20 kg molasses; 200 L water 1	3	Molasses availability
Urine acidification by sulphuric acid** addition	2.94 L concentrated H_2SO_4	10	Health risks during handling and treatment, potential agricultural negative impacts

*The additives used here are referred to a LAB solution obtained from sauerkraut, molasses and water, the prices are in accordance to those available on the market (Sancom, 2016).
**This refers to concentrated sulphuric acid, of not a high quality available on the market in Moldova (Kompas, 2017).

6.5.4 Potential for large scale sanitation and bio-waste management at schools with nutrient recycling to agriculture

According to recent ApaSan monitoring data (2014), only about 1.5 l of urine acummulates per capita per month and a school accumulates around 4 m^3 per year. The quantity seems to be insignificant, since 1.5 l is the quantity that one person produces per day. This means that the UDDTs are not used at their full potential and school children continue to use their old pit latrins as indicated for Slobozia Mare school (Figure 6.6 B). Estimations made during this research showed that a family of 3 people, who were adding lactic acid bacteria after each urination accumulated 600 l of urine per year or 0.55 l per person/day, including the LAB solution added during the toilet use, but also considering that the family members are at home only for a part of the day. These estimations indicate that a family of three people produce monthly 9 kg of faeces and cover material, 2 kg of toilet paper and 5 kg of kitchen waste or annually 108 kg, 24 kg and 57 kg respectively (Table 6.5).

Table 6.5. The quantity of different waste types estimated to be accumulated at household level and extrapolated for schools with UDDTs (mean±SD).

Accumulation rate	Faeces and cover material (kg)	Toilet paper (kg)	Kitchen waste (kg)	Urine (l)
Monthly	9 ± 3.6	2.0 ± 0.2	5 ± 0.1	
Annual	108 ± 43.3	24 ± 1.8	57 ± 1.4	600 ± 100
Annual per capita	36 ± 14.4	8 ± 0.6	19 ± 0.5	200 ± 33
Potentially to be generated per school (400 users, for 8 months)	14400	1000	5200	80000

Therefore, extrapolating these data for a school of 400 people, an amount of 14.4 tonnes of faeces and cover material, 1 tonne of toilet paper and 5 tons of kitchen waste can be obtained that can be processed via combined lactic acid fermentation and thermophilic composting (Table 6.5). The amount of the compost produced from 1 tonne of raw material is 150 kg. Considering the amount of faeces, toilet paper and bio-waste that could be generated (Quora, 2016), approximately 3000 tonnes of compost can be obtained during one school year. The price of potting soil in Moldova, depending of the package weight, ranges from 1.2 to 5 MDL l^{-1} (0.055-0.23 € l^{-1}). Thus, the profit obtained would be between 2400-15000 MDL (110-700 €), but it would be better if the schools use the compost for growing ornamental plants.

In addition, a school of 400 children produces about 80 tons of urine that could be lacto-fermented and used to fertilize a willow plantation. An urine tank is emptied usually once a year or once per two year, during which time the fertilizing value of urine is reduced due to ammonia volatilization during storage or application. However, only if the school children use the UDDTs and not the pit latrines. Chapter 5 showed that lacto-fermented urine is a better nitrogen source than stored urine and thus better maintaining its fertilizer value.

6.6 Potential entrepreneurial models

In a potential entrepreneurial model (Figure 6.8), excreta can be obtaind from UDDTs by a service provider. The entrepreneur treats the urine on-site (at school where the urine tanks are located) through the application of LAB supplemented with a source of carbohydrate. The urine, together with LAB and carbohydrate, is accumulated in tanks, which are emptied monthly and transported to a common short rotation copice (willow) plantation. The entrepreneur also collects once a month the accumulated quantity of faeces and bio-waste from the canteen and garden. Considering that willow requires 50-75 kgN/ha and that urine contains about 9 kgN/m^3 (Plămădeală *et al.*, 2011), the amount of nitrogen in the urine that accumulates in the UDDT during one school year (80 m^3) can fertilize 9 to 13 ha of willow. Taking into account the results from this research, in which 24-32 % of ammonia is prevented from being lost via volatilization by lactic acid fermentation (Chapter 5), the surface of irrigated willow could be increased by 20-30 %. However, this needs to be confirmed through field research. The willow can be used for producing renewable energy products (wood chips or pellets), in addition to the traditional Moldovan basket and furniture, which are highly valued in many countries. Also, willow cuttings easily set roots when planted, so reforestation with this tree species is simple (Figure 6.7).

Considering the price per ton of fresh willow woodchips for energy production, the total benefit could be up to 9000 euros per year, although the investment costs for willow planting, maintenance and harvesting along with the price of land would also need to be calculated to make any conclusions about the cost-benefit of this new sanitation approach. Schools can be encouraged to collect faeces, urine and vegetable waste through a discount card system for the collection services or for buying ornamental plants. More responsible schools can get small financial rewards that they could use to buy sanitary and school supplies. Wood from tree pruning as well as woodchips from the willows can be also used for local production of wood biochar. The compost can be used on site as growth medium for ornamental shrubs and the yard pruning waste for biochar production (Figure 6.7).

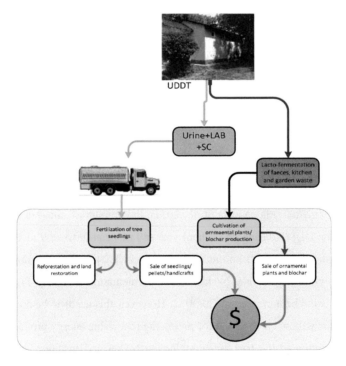

Figure 6.7 A potential entrepreneurial model where excreta is collected and treated via lactic acid fermentation to produce different revenue streams. LAB - lactic acid bacteria, SC - soluble carbohydrate source.

Depending on the type of plants grown, an optimal ratio of compost mixes can be determined. This research shows that a 1:3 compost:soil ratio was favourable for plant growth (Chapter 3). The profits obtained from growing ornamental plants depend on the plants species used as well as the demand for these plants. For example, decorative plants such as the fast growing shrub, *Hortensia,* are highly appreciated by gardeners for outdoor landscaping. Also *Petunia* is an annual plant, used widely in the gardens and for decoration of balconies. However, a more comprehensive cost-benefit analysis is required taking into account willow and ornamental plant growth, the price of compost at the local level, the demand for ornamental potted plants and shrubs and the amount of decorative plants produced.

6.7 Conclusions

The following main conclusions can be derived from this dissertation:

- The compost obtained via lactic acid fermentation of excreta, mixed with biochar is beneficial for agriculture. This type of compost produces a similar or even higher corn yield and improves soil quality compared to other fertilizers, such as NPK mineral fertilizers or simple stored urine, faeces and cattle manure.
- When lactic acid fermentation is combined with thermophilic composting with addition of biochar, it produces a safe soil conditioner with only passive aeration without the need for turning, this compost enhances plant germination and growth.
- Use of urine as a fertilizer can generate foul odours. When applying lactic acid fermentation, odour is largely controlled and there is less loss of nitrogen via ammonia volatilization.
- A cost-benefit analysis of the application of this technique is required. Also, additional field research on the effect on crops and soil is needed. Nonetheless, lactic acid fermentation of excreta has shown to be a sustainable alternative to mineral fertilizers which maintains soil fertility, where purchase of these fertilizers is not affordable.

References

Abel, S., Peters, A., Trinks, S., Schonsky, H., Facklam, M., Wessolek, G., 2013. Impact of biochar and hydrochar addition on water retention and water repellency of sandy soil. Geoderma 202, 183-191.

Abel-Aziz, S., Moustafa, Y., Hamed, H. Lactic acid bacteria in the green biocontrol against some phytopathogenic fungi: treatment of tomato seeds Journal of Basic and Applied Science Research, 4 (12), 1-9.

Abo-Amer, A.E., 2007. Characterization of a bacteriocin-like inhibitory substance produced by *Lactobacillus plantarum* isolated from Egyptian home-made yogurt. Science Asia 33, 313-319.

Agegnehu, G., Bird, M., Nelson, P., Bass, A., 2015. The ameliorating effects of biochar on soil quality and plant growth on a ferrasol. Soil res 53, 1-22.

Akpan-Idiok, A.U., Udo, I.A., Braide, E.I., 2012. The use of human urine as an organic fertilizer in the production of okra (*Abelmoschus esculentus*) in South Eastern Nigeria. Resour. Conserv. Recy. 62, 14-20.

Alattar, M.A., Alattar, F.N., Popa, R., 2016. Effects of micro-aerobic fermentation and black soldier fly larvae food scrap processing residues on the growth of corn plants (*Zea mays*). Plant Sci Today 3, 57-62.

Alattar, M.A., Green, T.R., Henry, J., Gulca, V., Tizazu, M., Bergstrom, R., Popa, R., 2012. Effect of Micro-aerobic Fermentation in Preprocessing Fibrous Lignocellulosic Materials. Applied Biochemistry and Biotechnology 167, 909-917.

Alburquerque, J.A., Salazar, P., Barrón, V., Torrent, J., del Campillo, M.d.C., Gallardo, A., Villar, R., 2013. Enhanced wheat yield by biochar addition under different mineral fertilization levels. Agronomy for Sustainable Development 33, 475-484.

Alidadi, H., Shamansouri, A.P.M., 2005. Combined compost and vermi-composting process in the treatment and bioconversion of sludge. Iranian Journal of Environmental Health Science & Engineering 2, 251-254.

Alidadi, H., Parvaresh, A., Shahmansouri, M., Pourmoghadas, H., Najafpoor, A., 2007. Combined compost and vermi-composting process in the treatment and bioconversion of sludge. Pakistan Journal of Biological Sciences: PJBS 10, 3944-3947.

Aliokin, O., Semionov, A., Skopintsev, B., 1973. Guidelines on chemical analysis of land waters (in Russian). Gydrometeoizdat, Leningrad.

Anand, C.K., Apul, D.S., 2014. Composting toilets as a sustainable alternative to urban sanitation - a review. Waste Manage 34, 329-343.

Andersson, R., Daeschel, M., Eriksson, C., 1988. Controlled lactic acid fermentation of vegetables. Proceed 8th Int Biotechnol Symp, Paris 1988/edited by G. Durand, L. Bobichon, J. Florent. Societe francaise de microbiologie, Paris, France.

Andreev, N., Ronteltap, M., Boincean, B., Lens, P.N.L., 2017. Treatment of source-separated human faeces via lactic acid fermentation combined with thermophilic composting. Compost Sci Util doi 10.1080/1065657X.2016.1277809.

Andreev, N., Ronteltap, M., Lens, P., Boincean, B., Bulat, L., Zubcov, E., 2016. Lacto-fermented mix of faeces and bio-waste supplemented by biochar improves the growth and yield of corn (*Zea mays* L.) Agriculture Ecosystems and Environment 232, 263-272.

Andreev, N., Ronteltap, M., Boincean, B., Lens, P., 2015. Processing of faeces from urine diverting dry toilets by combined biological processes: lactic acid fermentation and vermi/thermal composting. In: Mutnuri, S. (Ed.), 2nd International Conference on terra preta sanitation and decentralized wastewater system. BITS Pilani, Goa Campus, Goa, India, pp. 15-16.

Andreev, N., Ronteltap, M., Boincean, B., Lens, P., 2014. The effect of a terra preta-like soil improver on the germination and growth of radish and parsley. 1st International Terra preta sanitation conference. DBU, Hamburg, Germany.

Andreev, S., Andreev, N., 2010. Ecological sanitation. EcoSan concept. Dry toilets with separate excreta collection. The reuse of excreta in agriculture (in Romanian). WiSDOM, Chisinau.

Andries, S., 2006. The modification of humus content in Moldova chernozioms udner the process of explitation in agriculture. 18 th World Congress of Soil Sciences, Philadelphia, Pensylvania, USA.

Andries, S., 2007. Optimization of the nutritive regimes of the soils and productivity of crops (in Romanian). Pontos, Chisinau.

Andries, S., Cerbari, V., Filipciuc, V., 2014. The quality of Moldovan soils: issues and solutions. In: Dent, D. (Ed.), Soil as a world heritage. Springer, Dordrecht, pp. 9-13.

Andries, S., Leah, T., Povar, I., Lupascu, T., Filipciuc, V., 2013. Analysis of the use of fertilizers on different types of soil (in Romanian). Akademos 1, 123-131.

Anonymous, 2015. Nisporeni - Monthly weather history, 2013-2014. http://freemeteo.co.il/weather/nisporeni/history/monthly-history/?gid=617753&station=4627&month=8&year=2013&language=english&country=moldova

Anuradha, R., Suresh, A., Venkatesh, K., 1999. Simultaneous saccharification and fermentation of starch to lactic acid. Process Biochem 35, 367-375.

Arancon, N., Edwards, C.A., Webster, K.A., Buckerfield, J.C., 2010. The potential of vermicomposts as plant growth media for greenhouse crop production. Vermiculture technology: earthworms, organic wastes, and environmental management. CRC, Boca Raton, 103-128.

Asuming-Brempong, S., Nyalemegbe, K., 2014. The use of earthworms and biochar to mitigate increase in nitrous oxide production - a minireview. Global Advanced Research Journal of Agricultural Sciences 3, 035-041.

Atiyeh, R., Subler, S., Edwards, C., Bachman, G., Metzger, J., Shuster, W., 2000. Effects of vermicomposts and composts on plant growth in horticultural container media and soil. Pedobiologia 44, 579-590.

Atkinson, C.J., Fitzgerald, J.D., Hipps, N.A., 2010. Potential mechanisms for achieving agricultural benefits from biochar application to temperate soils: a review. Plant and Soil 337, 1-18.

Bagar, T., Kavčič, M., 2013. Humus: the forgoten answer to climate protection and sustainable farming. Conference VIVUS - environmentalism, agriculture, horticulture, food production and processing "Knowledge and experience for new enterpreneurial opportunities", Biotechnological Centre, Naklo, Slovenia, pp. 109-116.

Bai, Z., Dent, D., Olsson, L., Schaepman, M., 2008. Global assessment of land degradation and improvement 1: identification by remote sensing. Report 2008/01, FAO/ISRIC-Rome/Wageningen.

Baluhatîi, V., Costiuc, V., Vorobiov, V., Gojineţchi, I., 2002. Pedologic report. Soils of Boldureşti commune, Nisporeni district, their sustainable use. State Agency of Soil and Cadastre of the Republic of Moldova, Chisinau.

Barrena, R., Font, X., Gabarrell, X., Sanchez, A., 2014. Home composting versus industrial composting: Influence of composting system on compost quality with focus on compost stability. Waste Management, 34 (7), 1109-1116.

Batey, T., McKenzie, D., 2006. Soil compaction: identification directly in the field. Soil Use and Management 22, 123-131.

Beauchamp, E., 1988. Nitrogen transformations near urea in soil: effects of nitrification inhibition, nitrifier activity and liming. Fertility Research 18, 201-212.

Berezin, P., Gudima, I., 2002. Pore space in soil as living environment and basic factor in crop productivity (in Russian). Agriculture and ecology All Russian Institute of Agriculture and Erosion Control, Kursk, Russia.

Berg, W., Türk, M., Hellebrand, H., 2006. Effects of acidifying liquid cattle manure with nitric or lactic acid on gaseous emissions. Proceed Workshop Agric Air Qual: State Sci. Department of Communication Services, North Carolina State University, Potomac, Maryland, pp. 5-8.

Beriso, M., Otterpohl, R., 2013. Low-cost plant nutrients and organic matter recovery from human urine in terra preta sanitation TUHH, Website of Institute of Wastewater Management and Water Protection (TUHH).

Bernal, M.P., Alburquerque, J., Moral, R., 2009. Composting of animal manures and chemical criteria for compost maturity assessment. A review. Bioresource Technology 100, 5444-5453.

Bettendorf, T., Buzie, C., Glaser, B., Itchon, G., Klimek, F., Körner, I., Otterpohl, R., Pieplow, H., Schuetze, T., Wendland, C., Yemaneh, A., 2015. Terra Preta Sanitation 1 - Background, Principles and Innovations. Deutsche Bundesstiftung für Umwelt (DBU), Hamburg.

Bettendorf, T., Stoeckl, M., Otterpohl, R., 2014. Vermi-composting of municipal solid organic waste and faecal matter as part of Terra Preta Sanitation - a process and product assessment In: Bettendorf, T.W., C; Otterpohl, R. (Ed.), 1st International Conference on Terra preta sanitation. DBU, Hamburg, Germany.

Beauchamp, E. 1988. Nitrogen transformations near urea in soil: effects of nitrification inhibition, nitrifier activity and liming. Fertilizer Research, 18(3), 201-212.

Biederman, L.A., Harpole, W.S., 2013. Biochar and its effects on plant productivity and nutrient cycling: a meta-analysis. GCB Bioenerg 5, 202-214.

Blum, N., 2013. Soil and land resources for agricultural production: general trends and future scenarios - a worldwide perspective. International soil and water conservation research 1, 1-14.

Bodik, I., 2007. Current status of water supply and sanitation in the GWP CEE countries. In: Bodik, I., Ridderstolpe, P. (Eds.), Sustainable sanitation in Central and Eastern Europe - addressing the needs of small and medium-size settlements. GWP CEE Ljubljana, Slovenia.

Boincean, B., L., N., Stadnic, S., 2014. Productivity and fertility of the Balti chernozem under crop rotation with different systems of fertilzation. In: Dent, D. (Ed.), Soils as a world heritage. Springer, Dordrecht, the Netherlands, pp. 209-233.

Bottcher, J., Pieplow, H., Krieger, A., 2010. Method for the production of humus and nutrient-rich and water-storing soils or soil substrates for sustainable land use and development systems. In: States, U. (Ed.), Patent Application Publication, USA.

Böttger, S., Tows, I., Blecher, J., Kruger, M., H., S., Dorgeloh, E., Khan, P., Philipp, O., 2014. Applicability of terra preta produced from sewage sludge of decentralized wastewater systems in Germany. In: Bettendorf, T., Wendland, C., Otterpohl, R. (Eds.), 1st International Terra preta sanitation Conference. DOE, Hamburg, Germany.

Bowman, D.D., Liotta, J.L., McIntosh, M., Lucio-Forster, A., 2006. *Ascaris suum* egg inactivation and destruction by the vermi-composting worm, *Eisenia foetida*. Proceedings of the Water Environment Federation 2006, 11-18.

Bremner, J.M., Krogmeier, M.J. 1989. Evidence that the adverse effect of urea fertilizer on seed germination in soil is due to ammonia formed through hydrolysis of urea by soil urease. Proceedings of the National Academy of Sciences, 86(21), 8185-8188.

Brewer, C.E., Unger, R., Schmidt-Rohr, K., Brown, R.C., 2011. Criteria to select biochars for field studies based on biochar chemical properties. Bioenergy Research 4, 312-323.

Britto, D., Kronzucker, H., 2002. NH_4^+ toxicity in higher plants: a critical review. Journal of Plant Physiology 159, 567-584.

Bugaeva, T., Mironova, T., 2009. The meteorological and agrometeorological conditions of summer 2009. State Hydrometeorological Service of Moldova, www.meteo.md/rus/leto09.htm.

Bulbo, M., Yemaneh, A., Amlaku, T., Otterpohl, R., 2014. Assessment of availability of terra preta sanitation precursors in Arba Minch, Ethiopia. In: Bettendorf, T., Wendland, C., Otterpohl, R. (Eds.), 1st International Conference on Terra preta sanitation. DBU, Hamburg, Germany.

Burge, W., Colacicco, D., Cramer, W., Epstein, E., 1978. Criteria for control of pathogens during sewage sludge composting. National Conference on Design of Municipal Sludge Compost Facilities, Chicago, pp. 124-129.

Burgos, P., Madejón, E., Cabrera, F., 2006. Nitrogen mineralization and nitrate leaching of a sandy soil amended with different organic wastes. Waste Management Resources 24, 175-182.

Burns, L., Stevens, R., Smith, R., Cooper, J., 1995. The occurrence and possible sources of nitrite in a grazed, fertilized, grassland soil. Soil Biology and Biochemistry 27, 47-59.

Buss, W., Mašek, O., 2014. Mobile organic compounds in biochar – a potential source of contamination – phytotoxic effects on cress seed (*Lepidium sativum*) germination. Journal of Environmental Management 137, 111-119.

Buzie-Fru, C.A., 2010. Development of a continuous single chamber vermi-composting toilet with urine diversion for on-site application. Technical University Hamburg-Harburg. Universitätsbibliothek, Hamburg, p. 136.

Buzie, C., Körner, I., 2014. Chapter IV: Composting of bioresources for Terra preta-inspired products. In: Bettendorf, T., Wendland, C., Otterpohl, R. (Eds.), Terra Preta Sanitation. Background, principles and innovation. DBU, Hamburg, Germany, pp. 86-119.

Caballero-Hernández, A., Castrejón-Pineda, F., Martınez-Gamba, R., Angeles-Campos, S., Pérez-Rojas, M., Buntinx, S., 2004. Survival and viability of *Ascaris suum* and *Oesophagostomum dentatum* in ensiled swine faeces. Bioresource Technology 94, 137-142.

Canfield, J., Goldner, B., 1964. Research on applied bioelectrochemistry. NASA Technical Report Magna Corporation, Anaheim 127.

Carlson, C., Reckinger, N., 2016. Opportunities in recycling phosphorus. In: Inc, F.I. (Ed.), http://feeco.com/opportunities-in-recycling-phosphorus/.

Carvalho, I., Detmann, E., Mantovani, H., Paulino, M., Valadares Filho, S., Costa, V., Gomes, D., 2011. Growth and antimicrobial activity of lactic acid bacteria from rumen fluid according to energy or nitrogen source. Revue Brasilian Zootechny 40, 1260-1265.

Caswell, L., Webb, K., Fontenot, J., 1977. Fermentation, nitrogen utilization. Digestibility and palatability of broiler litter ensiled with high moisture corn grain. Journal of Animal Science 44, 803-813.

Chaoui, H.I., Zibilske, L.M., Ohno, T., 2003. Effects of earthworm casts and compost on soil microbial activity and plant nutrient availability. Soil Biology and Biochemistry 35, 295-302.

Cheng, C.H., Lehmann, J., Thies, J., Burton, S., Engelhard, M., 2006. Oxidation of black carbon by biotic and abiotic processes. Organic Geochemistry 37, 1477-1488.

Cho, K.M., Math, R.K., Islam, S.M.A., Lim, W.J., Hong, S.Y., Kim, J.M., Yun, M.G., Cho, J.J., Yun, H.D., 2009. Biodegradation of chlorpyrifos by lactic acid bacteria during kimchi fermentation. Journal of Agricultural and Food Chemistry 57, 1882-1889.

Cho, R., 2013. Phosphorus: essential to life - are we running out? In: University, C. (Ed.), Agriculture, Earth Sciences. Earth Institute, Columbia.

Chowdhury, M.A., de Neergaard, A., Jensen, L.S., 2014. Potential of aeration flow rate and bio-char addition to reduce greenhouse gas and ammonia emissions during manure composting. Chemosphere 97, 16-25.

Ciari, D., Manning, D., Allaure, A., 2015. Historical and technical developments of potassium resources. Science of the Total Environment 502, 590-601.

Cirja, M., Ivashechkin, P., Schäffer, A., Corvini, P.F., 2008. Factors affecting the removal of organic micropollutants from wastewater in conventional treatment plants (CTP) and membrane bioreactors (MBR). Reviews in Environ Science and Biotechnology 7, 61-78.

Cordell, D., White, S., 2011. Peak phosphorus: clarifying the key issues of a vigorous debate about long-term phosphorus security. Sustainability 3, 2027-2049.

Cotta, M.A., Whitehead, T.R., Zeltwanger, R.L., 2003. Isolation, characterization and comparison of bacteria from swine faeces and manure storage pits. Environmental Microbiology 5, 737-745.

Dal Bello, F., Walter, J., Hammes, W., Hertel, C., 2003. Increased complexity of the species composition of lactic acid bacteria in human feces revealed by alternative incubation condition. Microbial Ecology 45, 455-463.

Date, J., 1958. A quantitative method for estimation of some carbohydrates on filter paper. Scandinavian Journal of Clinical Laboratory Investigations 10, 149-154.

Dawson, C., Hilton, J., 2011. Fertilizer availability in a resource limited world production and recycling of nitrogen and phosphorus. Food Policy 36, 14-22.

DeLuca, T., MacKenzie, D., Gundale, M., 2009. Biochar effects on soil nutrient transformations. In: Lehmann, J., Joseph, S. (Eds.), Biochar for environmental management. science and technology. Earthscan, London, pp. 251-265.

Devi, S.H., Vijayalakshmi, K., Jyotsna, K.P., Shaheen, S., Jyothi, K., Rani, M.S., 2009. Comparative assessment in enzyme activities and microbial populations during normal and vermi-composting. Journal of environmental Biology 30, 1013-1017.

Dias, B., Silva, C., Higashikawa, F., Roig, A., Sanches, M., 2010. Use of biochar as a bulking agent for the composting of poultry manure: effect of organic matter degradation and humification. Bioresource Technology 101, 1239-1246.

Domeizel, M., Khalil, A., Prudent, P., 2004. UV spectroscopy: a tool for monitoring humification and for proposing an index of the maturity of compost. Bioresource Technology 94, 177-184.

Dospehov, B., 1985. Technique of field experience (in Russian). Agropromizdat, Moskow.

Dowd, S.E., Callaway, T.R., Wolcott, R.D., Sun, Y., McKeehan, T., Hagevoort, R.G., Edrington, T.S., 2008. Evaluation of the bacterial diversity in the feces of cattle using 16S rDNA bacterial tag-encoded FLX amplicon pyrosequencing (bTEFAP). BMC Microbiology 8, 125.

Downie, A., Van Zwieten, L., Smernik, R., Morris, S., Munroe, P., 2011. Terra preta australis: reassessing the carbon storage capacity of temperate soils. Agriculture Ecosystems and Environment 140, 137-147.

Dumbrepatil, A., Adsul, M., Chaudhari, S., Khire, J., Gokhale, D., 2008. Utilization of molasses sugar for lactic acid production by *Lactobacillus delbrueckii* subsp.*delbrueckii* mutant Uc-3 in batch fermentation. Applied Environmental Microbiology 74, 333-335.

Dunn, M.S., Shankman, S., Camien, M.N., Block, H., 1947. The amino acid requirements of twenty-three lactic acid bacteria. Journal of Biological Chemistry 168, 1-22.

Eastman, B.R., Kane, P.N., Edwards, C.A., Trytek, L., Gunadi, B., Stermer, A.L., Mobley, J.R., 2001. The effectiveness of vermiculture in human pathogen reduction for USEPA biosolids stabilization. Compost Science and Utilization 9, 38-49.

EBC (2012) 'European Biochar Certificate - Guidelines for a Sustainable Production of Biochar.'European Biochar Foundation (EBC), Arbaz, Switzerland.

http://www.europeanbiochar.org/en/download. Version 6.2E of 04th February 2016, DOI: 10.13140/RG.2.1.4658.7043

Eckmeier, E., Gerlach, R., Skjemstad, J., Ehrmann, O., Schmidt, M., 2007. Minor changes in soil organic carbon and charcoal concentrations detected in a temperate deciduous forest a year after an experimental slash-and-burn. Biogeoscience 4, 377-383.

Edwards, C., Burrows, I., Fletcher, K., 1984. The use of earthworms for composting farm wastes. In: Gasser, J. (Ed.), Composting of agricultural and other wastes. Elsevier Applied Science Publishers, London.

Elad, Y., David, D.R., Harel, Y.M., Borenshtein, M., Kalifa, H.B., Silber, A., Graber, E.R., 2010. Induction of systemic resistance in plants by biochar, a soil-applied carbon sequestering agent. Phytopathology 100, 913-921.

Elizarova, O.V., 2000. Bichromate oxidation as a method of characterization of water quality. Moskow Pedagogical Institute, Moskow.

Enders, A., Hanley, K., Whitman, T., Joseph, S., Lehmann, J., 2012. Characterization of biochars to evaluate recalcitrance and agronomic performance. Bioresource Technology 114, 644-653.

Epstein, E., Wilson, G.B., Byrge, W.D., Mullen, D.C., Enkiri, N.K., 1976. A force aeration system for composting wastewater sludge. Journal of the Water Pollution Control Federation 48, 688-694.

Erickson, C., 2003. Historical ecology and future explorations. Amazonian Dark Earths: origin, properties, management., 455-500.

Erickson, L.E., Fayet, E., Davis, L.C., 2004. Lactic Acid Fermentation. In: National Agricultural Biosecurity Center Consortium (Ed.), Carcass Disposal: A Comprehensive Review. Kansas State University, Kansas.

Factura, H., Bettendorf, T., Buzie, C., Pieplow, H., Reckin, J., Otterpohl, R., 2010. Terra Preta sanitation: re-discovered from an ancient Amazonian civilisation - integrating sanitation, bio-waste management and agriculture. Water Science and Technology 61, 2673-2679.

FAO, 2011. The state of the world's land and water resources for food and agriculture (SOLAW) - managing systems at risk. Earthscan Rome and London.

Fhoula, I., Najjari, A., Turki, Y., Jaballah, S., Boudabous, A., Ouzari, H., 2013. Diversity and antimicrobial properties of lactic acid bacteria isolated from rhizosphere of olive trees and desert truffles of Tunisia. Biological and Medical Research International.

Fornes, F., Mendoza-Hernández, D., García-de-la-Fuente, R., Abad, M., Belda, R.M., 2012. Composting versus vermi-composting: a comparative study of organic matter evolution through straight and combined processes. Bioresource Technology 118, 296-305.

Frausin, V., Fraser, J.A., Narmah, W., Lahai, M., Winnebah, T., Fairhead, J., Leach, M., 2014. "God Made the Soil, but We Made It Fertile": Gender, Knowledge, and Practice in the Formation and Use of African Dark Earths in Liberia and Sierra Leone Human Ecology 42, 695-710.

Frazier, W., 1967. Food microbiology. McGraw Hill Company, New York.

Frederickson, J., Howell, G., 2003. Large-scale vermi-composting: emission of nitrous oxide and effects of temperature on earthworm populations: the 7th international symposium on earthworm ecology· Cardiff· Wales· 2002. Pedobiology 47, 724-730.

Frederickson, J., Knight, D., 1988. The use of anaerobically digested cattle solids for vermiculture. In: Edwards, C.A., Neuhauser, E.F. (Eds.), Earthworms in Waste and Environmental Management. SPB Academic Publishing, The Hague, pp. 33-47.

Freitag, D.G., Meihoefer, H., 2000. The use of Effective Microorganisms (EM) in organic waste management. Effective Microorganisms@ emtrading. com.

Frost, J., Stevens, R., Laughlin, R., 1990. Effect of separation and acidification of cattle slurry on ammonia volatilization and on the efficiency of slurry nitrogen for herbage production. Journal of Agricultural Science 115, 49-56.

Fu, M., 1989. Effect of pH and organic acids on nitrogen transformations and metal dissolution in soils. Iowa State University, Iowa.

Gajić, A., Koch, H.-J., 2012. Sugar beet (L.) growth reduction caused by hydrochar is related to nitrogen supply. Journal of Environmental Quality 41, 1067-1075.

George, C., Wagner, M., Kücke, M., Rillig, M.C., 2012. Divergent consequences of hydrochar in the plant–soil system: arbuscular mycorrhiza, nodulation, plant growth and soil aggregation effects. Applied Soil Ecology 59, 68-72.

Giani, L., Makowsky, L., Mueller, K., 2014. Plaggic Anthrosol: Soil of the Year 2013 in Germany: An overview on its formation, distribution, classification, soil function and threats. Journal of Plant Nutrition and Soil Science 177, 320-329.

Gibbs, R.A., Hu, C.J., Ho, G.E., Unkovich, I., 1997. Regrowth of faecal coliforms and salmonellae in stored biosolids and soil amended with biosolids. Water Science & Technology 35, 269-275.

Glaser, B., 2007. Prehistorically modified soils of central Amazonia: a model for sustainable agriculture in the twenty-first century. Philosophical Transactions of the Royal Society B: Biological Sciences 362, 187.

Glaser, B., 2015. Biochar as soil amendment: facts and myths. In: TUHH, A., GFEU e.v., WECF (Ed.), Terra preta sanitation 1: background, principles and innovation. Deutsche Bundesstiftung Umwelt, Hamburg, pp. 16-29.

Glaser, B., Birk, J.J., 2012. State of the scientific knowledge on properties and genesis of Anthropogenic Dark Earths in Central Amazonia (terra preta de Índio). Geochimica et Cosmochimica Acta 82, 39-51.

Gleick, P.H., 2003. Global freshwater resources: soft-path solutions for the 21st century. Sci 302, 1524-1528.

Glibert, P., Harrison, J., Heil, C., Seitzinger, S., 2006. Escalating worldwide use of urea – a global change contributing to coastal eutrophication. Biogeochemistry 77, 441-463.

Gómez-Brandón, M., Lazcano, C., Domínguez, J., 2008. The evaluation of stability and maturity during the composting of cattle manure. Chemosphere 70, 436-444.

Gorman, S., 2011. Lacto-fermented vegetables and their potential in the developing world. Food Chain 1, 87-94.

GOST UJCN 13634-90, 2010. Corn. Requirements for state purchases and deliveries. Standard Inform, Moskow.

GOST 26489-85, 1985. Determination of exhangeable ammonium by CINAO method. Soils. State Standard Soiuz SSR, Moskow.

GOST 27894.4-88, 1989. Peat and products of its processing for agriculture. Methods of determination of nitrate nitrogen. State Standard Soiuz SSR, Moskow.

Gotaas, H., 1956. Composting: sanitatry disposal and reclamation of organic wastes. WHO, Geneva.

Graham, J.P., Polizzotto, M.L., 2013. Pit latrines and their impacts on groundwater quality: a systematic review. Environtal Health Perspectives 121.

Green, T.R., Popa, R., 2011. Turnover of carbohydrate-rich vegetal matter during micro-aerobic composting and after amendment in soil. Applied Biochemical Biotechnology 165, 270-278.

Greg, C., Sing, K., 1984. Adsorption. Specific surface area. Porosity (in Russian). Mir, Moskow.

Güzel-Seydim, Z., Seydim, A., Greene, A., Bodine, A., 2000. Determination of organic acids and volatile flavor substances in kefir during fermentation. Journal of Food composition and Analysis 13, 35-43.

Guzha, E., Nhapi, I., Rockstrom, J., 2005. An assessment of the effect of human faeces and urine on maize production and water productivity. Physics and Chemistry of the Earth, Parts A/B/C 30, 840-845.

Hamer, U., Marschner, B., 2005. Priming effects in different soil types induced by fructose, alanine, oxalic acid and catechol additions. Soil Biological Biochemistry 37, 445-454.

Hanjra, M.A., Qureshi, M.E., 2010. Global water crisis and future food security in an era of climate change. Food Policy 35, 365-377.

Hartung, J., Phillips, V., 1994. Control of gaseous emissions from livestock buildings and manure stores. Journal of Agricultural Engineering Research. 57, 173-189.

Hassan, F., El-Gawad, M., Enab, A., 2012. Flavour compounds in cheese (review). Journal of Academic Research 4, 169-181.

Hata, K., 1982. Cultivation of deodorizing *Lactobacillus* strain, storage, thereof, and composition containing living cells thereof. In: US Patents (Ed.), USA.

Hatfield, J., Sauer, T., Pruger, J., 2014. Managing soils to achieve greater water use efficiency: a review. Publication from USDA-ARS/UNL Faculty Paper 1341.

Hawke, B.G., Baldock, J.A., 2010. Ammonia volatilisation from urea fertilizer products applied to an alkaline soil. 19th World Congress of Soil Science.

Hayes, W., Randle, P., 1968. Use of molasses as an ingredient of wheat straw mixtures used for the preparation of mushroom composts. Annual Report Glasshouse Crops Research Institute 1968, pp. 142-147.

Haynes, R., Williams, P., 1993. Nutrient cycling and soil fertility in the grazed pasture ecosystem. Advances in Agronomy (USA).

Heinonen-Tanski, H., Sjöblom, A., Fabritius, H., Karinen, P., 2007. Pure human urine is a good fertilizer for cucumbers. Bioresource Technology 98, 214-217.

Hellström, D., Johansson, E., Grennberg, K., 1999. Storage of human urine: acidification as a method to inhibit decomposition of urea. Ecological Engineering 12, 253-269.

Hijikata, N., Yamauchi, N., Ishiguro, M., Ushijima, K., Funamizu, N., 2015. Suitability of biochar as a matrix for improving the performance of composting toilets. Waste Management Resources 33, 313-321.

Haug, R.T., 1993. The practical handbook of compost engineering. Lewis Publishers, Boca Raton.

Hayes, W., Randle, P., 1968. Use of molasses as an ingredient of wheat straw mixtures used for the preparation of mushroom composts. Annual Report Glasshouse Crops Research Institute 1968, pp. 142-147.

Hoda, A., Yomna, A., Shadia, M. 2011. In vivo efficacy of lactic acid bacteria in biological control against *Fusarium oxysporum* for protection of tomato plant. Life Science Journal, 8(4), 463-468.

Hotta, S., Funamizu, N., 2007. Biodegradability of faecal nitrogen in composting process. Bioresource Technology 98, 3412-3414.

Hill, G.B., Baldwin, S.A., 2012. Vermi-composting toilets, an alternative to latrine style microbial composting toilets, prove far superior in mass reduction, pathogen destruction, compost quality, and operational cost. Waste Management 32, 1811-1820.

Hill, G.B., Lalander, C., Baldwin, S.A., 2013. The Effectiveness and Safety of Vermi-Versus Conventional Composting of Human Feces with *Ascaris suum* Ova as Model Helminthic Parasites. Journal of Sustainable Development 6 (4), http://www.ccsenet.org/jsd.

Hitman, A., Bos, K., Bosch, M., Van der Kolk, A., 2013. Fermentation versus composting. Feed inovation Services, Wageningen, The Netherlands.

Hoda, A., Yomna, A., Shadia, M., 2011. In vivo efficacy of lactic acid bacteria in biological control against *Fusarium oxysporum* for protection of tomato plant. Life Science Journal 8, 463-468.

Hoff, H., 2011. Understanding the nexus: Background paper for the Bonn 2011 Nexus Conference: the water, energy and food security nexus. Nexus Conference: the water, energy and food security nexus. Stockholm Environment Institute, Bonn.

Hogg, D. Favoino, E, Nielsen, N. Thompson, J., Wood, K. Penschke, A. Economides, D., Papageorgiou S., 2002. Economic analysis of options for managing biodegradable municipal waste. Final Report to the European Commission, Eunomia Research and Consulting, Bristol.

Höglund, C., Stenström, T.A., Ashbolt, N., 2002. Microbial risk assessment of source-separated urine used in agriculture. Waste Management Resources 20, 150-161.

Hotta, S., Funamizu, N., 2007. Biodegradability of faecal nitrogen in composting process. Bioresource Technology 98, 3412-3414.

Hui-Lian, X., Wang, R., Mridha, A.U., 2001. Effects of Organic Fertilizers and a Microbial Inoculant on Leaf Photosynthesis and Fruit Yield and Quality of Tomato Plants. Journal of Crop Production 3, 173-182.

Hurtado, D., 2005. Compost latrines in rural panama: Design, construction and evaluation of pathogen removal. Michigan Technological University.

Ikawa, M., Snell, E., 1960. Cell wall composition of lactic acid bacteria. Journal of Biological Chemistry 235, 1376-1382.

ISO, 1998. Microbiology of food and animal feeding stuffs. Horizontal method for the enumeration of mesophilic lactic acid bacteria. Colony-count technique at 30 °C. The International Organization for Standardization.

Jaatinen, S., Palmroth, M., Rintala, J., Tuhkanen, T., 2016. The Effect of Urine Storage on Antiviral and Antibiotic Compounds in the Liquid Phase of Source Separated Urine. Environmental Technology 37, 2189-2198.

Javed, M.M., Zahoor, S., Shafaat, S., Mehmooda, I., Gul, A., Rasheed, H., Bukhari, S.A.I., Aftab, M.N., 2012. Wheat bran as a brown gold: Nutritious value and its biotechnological applications. African Journal of Microbiological Research 6, 724-733.

Jawad, A.H., Alkarkhi, A.F., Jason, O.C., Easa, A.M., Norulaini, N.N., 2013. Production of the lactic acid from mango peel waste - Factorial experiment. Journal of King Saud University Science 25, 39-45.

Jianyao, C., Ya, W., Hongbo, Z., Xinfeng, Z., 2010. Overview on the studies of nitrate pollution in groundwater. Progress in Geography 25, 34-44.

Jiménez, B., Drechsel, P., Koné, D., Bahri, A., Raschid-Sally, L., Qadir, M., 2010. Wastewater, sludge and excreta use in developing countries: an overview. In: Drechsel, P., Scott, C., Raschid-Sally, L., Redwood, M., Babri, A. (Eds.), Wastewater Irrigation and health. Assessing and mitigating risk in low-income countries. IWMI, Ottawa, pp. 3-29.

Jiménez, E.I., Garcia, V.P., 1989. Evaluation of city refuse compost maturity: a review. Biological wastes 27, 115-142.

Jindo, K., Sonoki, T., Matsumoto, K., Canellas, L., Roig, A., Sanchez-Monedero, M.A., 2016. Influence of biochar addition on the humic substances of composting manures. Waste Management, 49, 545-552.

Jindo, K., Suto, K., Matsumoto, K., García, C., Sonoki, T., Sanchez-Monedero, M.A., 2012. Chemical and biochemical characterisation of biochar-blended composts prepared from poultry manure. Bioresource Technology 110, 396-404.

Johansson, M., Jönsson, H., Höglund, C., Richert-Stintzin, A., Rodhe, L., 2000. Urine Separation, Closing the Nutrient Cycle. Final Report On The R&D Project Source-separated human urine, a future source of fertilizer for agriculture in the Stockholm region. Stockholm Water Company.

Johnson, M., Askin, D., 2005. Container grown experiment. AusAid, NARI.

Johnson, M., Hilbert, I., Jollymore, A., 2012. Biochar as a substitute for peat in greenhouse growing media: soil water characteristics and carbon leaching dynamics. AGU Fall Meeting Abstracts, p. 0543.

Jones, A., Panagos, P., Barcelo, S., Bouraoui, F., Bosco, C., Dewitte, O., Gardi, C., Erhard, M., Hervas, J., Hiederer, R., Jefferry, S., Lukewille., A., 2012. The state of soil in Europe. In: Centre, J.R. (Ed.), JRC Reference Report, Louxemburg.

Jönsson, H., 2003. The role of ecosan in achieving sustainable nutrient cycles. IWA 2nd International Symposium on Ecological Sanitation, pp. 35-40.

Jönsson, H., Stintzing, A.R., Vinnerås, B., Salomon, E., 2004. Guidelines on the use of urine and faeces in crop production. EcoSanRes Programme.

Juarez, P.D.A., Lugo de la Fuente, J., Vaca Paulin, R., 2011. Vermicompost in the process of organic waste and sewage sludge in the soil. Tropical and Subtropical Agroecosystems 14.

Jindo, K., Suto, K., Matsumoto, K., García, C., Sonoki, T., Sanchez-Monedero, M.A., 2012. Chemical and biochemical characterisation of biochar-blended composts prepared from poultry manure. Bioresour Technol 110, 396-404.

Juris, P., Rataj, D., Ilavska, I., Vasilkova, Z., 1997. Survival of model helminth eggs and larvae (*Ascaris suum*, *Oesophagostomum sp.*) in the ensilaging process. Veterinary Medicina 42, 165-169.

Jusoh, M.L.C., Manaf, L.A., Latiff, P.A., 2013. Composting of rice straw with effective microorganisms (EM) and its influence on compost quality. Iranian Journal of Environmental Health Science & Engineering 10, 1.

Kaiser, J., 2006. An analysis of the use of desiccant as a method of pathogen removal in compost latrines in rural panama. Michigan Technological University.

Kammann, C.I., Schmidt, H.P., Messerschmidt, N., Linsel, S., Steffens, D., Müller, C., Koyro, H.-W., Conte, P., Stephen, J., 2015. Plant growth improvement mediated by nitrate capture in co-composted biochar. Scientific Reports 5.

Kämpf, N., Woods, W., Sombroek, W., Kern, D., Cunha, T., 2004. Classification of Amazonian Dark Earths and other ancient anthropic soils. In: Lehman, D.S., Kern, D., Glaser, B., Woods, W. (Eds.), Amazonian Dark Earths: origin, properties, management. Kluwer Academic Publisher, Dordrecht, pp. 77-102.

Kamra, D., Jakhmola, R., Singh, R., 1984. Ensiling of cattle waste and wheat straw with or without green maize. Agricultural Wastes 9, 309-313.

Kang, D., Oh, H., Ham, J., Kim, J., Yoon, C., Ahn, Y., Kim, H., 2005. Identification and characterization of hydrogen peroxide-generating *Lactobacillus fermentum* CS12-1. Asian-Australas. Journal of Animal Science, 18, 90-95.

Kantha, T., Chaiyasut, C., Sukrong, D.K.S., Muangprom, A., 2012. Synergistic growth of lactic acid bacteria and photosynthetic bacteria for possible use as a bio-fertilizer. African Journal of Microbiological Research 6, 504-511.

Karsten, G.R., Drake, H.L., 1997. Denitrifying bacteria in the earthworm gastrointestinal tract and in vivo emission of nitrous oxide (N_2O) by earthworms. Applied Environmental Microbiology 63, 1878-1882.

Katukiza, A., Ronteltap, M., Niwagaba, C., Foppen, J., Kansiime, F., Lens, P., 2012. Sustainable sanitation technology options for urban slums. Biotechnological Advances 30, 964-978.

Kawa, N., Oyuela-Caycedo, A., 2008. Amazonian dark earth: a model of sustainable agriculture of the past and future? The international Journal of Environmental, Cultural, Economic and Social Sustainability 4, 9-16.

Kimani-Murage, E.W., Ngindu, A.M., 2007. Quality of water the slum dwellers use: the case of a Kenyan slum. Journal of Urban Health 84, 829-838.

Kirchmann, H., Pettersson, S., 1994. Human urine-chemical composition and fertilizer use efficiency. Nutrient Cycling in Agroecosystems 40, 149-154.

Knowles, O., Robinson, B., Contangelo, A., Clucas, L., 2011. Biochar for the mitigation of nitrate leaching from soil amended with biosolids. Science of Total Environment 409, 3206-3210.

Koh, R.H., Song, H.G. 2007. Effects of application of *Rhodopseudomonas* sp. on seed germination and growth of tomato under axenic conditions. Journal of Microbiology and Biotechnology, 17 (11), 1805-1810.

Kompas, 2016. Companies, acid sulfuric. Chisinau. Available from http://md.kompass.com/c/ecochimie-srl/md007562/. Last visit on 03.02.2017.

Kone, D., 2010. Making urban excreta and wastewater management contribute to cities economic development: a pradigm shift Water Policy 12, 602-610.

Kosseva, M., 2013. Food industry wastes assessment and opportunities. In: Kosseva, M., Webb, C. (Eds.), Food industry wastes assessment and recuperation. Academic Press, London, UK, pp. 3-14.

Krajewska, B., van Eldik, R., Brindell, M., 2012. Temperature-and pressure-dependent stopped-flow kinetic studies of jack bean urease. Implications for the catalytic mechanism. JBIC Journal of Biological and Inorganic Chemistry 17, 1123-1134.

Krause, A., Kaupenjohann, M., George, E., Koeppel, J., 2015a. Nutrient recycling from sanitation and energy systems to the agroecosystem-Ecological research on case studies in Karagwe, Tanzania. African Journal of Agricultural Research 10, 4039-4052.

Krause, A., Nehls, T., George, E., Kaupenjohann, M., 2015b. Organic waste from bioenergy and ecological sanitation as soil fertility improver: a field experiment in a tropical andosol. Soil Discussions 2, 1221-1261.

Krupenikov, I., 1967. The influence of soils on yield and quality of tobacco (review) (in Russian). Moldavian Research Institute, Kishinev.

Krupenikov, I.A., Boincean, B.P., Dent, D., 2011a. Humus – Guardian of Fertility and Global Carbon Sink. The Black Earth. Springer, pp. 39-50.

Krupenikov, I.A., Boincean, B.P., Dent, D., 2011b. The Past, Present and Future of the Chernozem. Earth and Environmental Science 10, 131-138.

Kunin, C.M., Hua, T.H., White, L.V.A., Villarejo, M., 1992. Growth of *Escherichia coli* in human urine: role of salt tolerance and accumulation of glycine betaine. Journal of Infectious Diseases, 166, 1311-1315.

Kuromiya, S., Saotome, O., Yamaguchi, I., Habu, K., 2010. Method for producing lactic acid ester. In: Office, U.S.P. (Ed.), Cynwyd, Pensilvania, USA.

Kvarnström, E., Emilsson, K., Stintzing, A.R., Johansson, M., Jönsson, H., af Petersens, E., Schönning, C., Christensen, J., Hellström, D., Qvarnström, L., 2006. Urine diversion: one step towards sustainable sanitation. EcoSanRes Programme, Stockholm.

Lal, R., 2002. Soil and carbon dynamics in cropland and rangeland. Environmental Pollution 116, 353-362.

Lal, R., 2009. Challenges and opportunities in soil organic matter research. European Journal of Soil Science 60, 158-169.

Langergraber, G., Muellegger, E., 2005. Ecological sanitation - a way to solve global sanitation problems? Environmental International 31, 433-444.

Larsen, T.A., Alder, A.C., Eggen, R.I., Maurer, M., Lienert, J., 2009. Source Separation: Will We See a Paradigm Shift in Wastewater Handling? Environmental Science and Technology 43, 6121-6125.

Larson, A., Kallio, R., 1954. Purification and properties of bacterial urease. Journal of Bacteriology 68, 67.

Lawal, H., Girei, H., 2013. Infiltration and organic carbon pools under the long term use of farm yard manure and mineral fertilizer. International Journal of Advanced Agricultural Research 1, 92-101.

Lawton, K., 2014. Influence of humic substances on soil health, fertilizer and water use efficiency. Far West Winter Agribusiness Conference. University of Idaho, Twin Falls, Idaho.

Lazcano, C., Arnold, J., Tato, A., Zaller, J., Domínguez, J., 2009. Compost and vermicompost as nursery pot components: effects on tomato plant growth and morphology. Spanish Journal of Agricultural Research, 944-951.

Leah, T., 2012. Land resources management and soil degradation factors in the Republic of Moldova. The 3rd International Symposium, "Agrarian Economy and Rural Development - realities and perspectives for Romania", Bucharest, Romania,.

Leah, T., Andries, S., 2012. The effectiveness of fertilizers on nutrient balance in terms of soil degradation in Republic of Moldova. Annals of the Academy of Romanian Scientists, Series on Agriculture, Silviculture and Veterinary Medicine Sciences 1, 48-61.

Lee, E., Lee, H., Kim, Y., Sohn, K., Lee, K., 2011. Hydrogen peroxide interference in chemical oxygen demand during ozone based advanced oxidation of anaerobically digested livestock wastewater. International Journal of Environmental Science and Technology 8, 381-388.

Lehmann, J., Gaunt, J., Rondon, M., 2006. Biochar sequestration in terrestrial ecosystems–a review. Mitigation and adaptation strategies for global change 11, 395-419.

Lehmann, J., J., S., Steiner, C., Nehls, T., 2003. Nutrient availability and leaching in an archeological Anthrosol and a Ferrasol of the Central Amazon basin: fertilizer, manure and charcoal amendments. Plant and Soil 249, 343-357.

Lehmann, J., Joseph, S., 2009. Biochar for environmental management: an introduction. In: Lehmann, J., Joseph, S. (Eds.), Biochar for environmental managment science and technology. Earthscan, London, Sterling, pp. 1-9.

Leng, R., 1984. The potential of solidified molasses-based blocks for the correction of multinutritional deficiencies in buffaloes and other ruminants fed low-quality agro-industrial byproducts. The use of nuclear techniques to improve domestic buffalo production in Asia.

Lentz, M., D., R., Lehrsch, G.A., 2012. Net nitrogen mineralization from past years manure and fertilizer applications. Soil Sci Society of America Journal 76, 1005-1015.

Liang, B., Lehmann, J., Solomon, D., Kinyangi, J., Grossman, J., O'neill, B., Skjemstad, J., Thies, J., Luizao, F., Petersen, J., 2006a. Black carbon increases cation exchange capacity in soils. Soil Science Society of America Journal 70, 1719-1730.

Liang, Y., Leonard, J., Feddes, J., McGill, W., 2006b. Influence of carbon and buffer amendment on ammonia volatilization in composting. Bioresource Technology 97, 748-761.

Lievens, C., Mourant, D., Gunawan, R., Hu, X., Wang, Y., 2015. Organic compounds leached from fast pyrolysis mallee leaf and bark biochars. Chemosphere 139, 659-664.

Lim, T., Pak, T., Jong, C., 1999. Yields of rice and maize as affected by effective microorganisms. In: Centre, I.N.F.R. (Ed.), Proceedings of the 5th International Conference on Kyusei Nature Farming and Effective Microorganisms for Agricultural and Environmental Sustainability, Bangkok, Thailand, pp. 92-98.

Limanska, N., Korotaeva, N., Biscola, V., Ivanytsia, T., Merlich, A., Franco, B., Chobert, J., Ivanytsia, V., Haertlé, T., 2015. Study of the Potential Application of Lactic Acid Bacteria in the Control of Infection Caused by *Agrobacterium tumefaciens*. Plant Pathology and Microbiology 6.doi:10.4172/2157-7471.1000292

Lind, B.-B., Ban, Z., Bydén, S., 2001. Volume reduction and concentration of nutrients in human urine. Ecological Engineering 16, 561-566.

Liu, B., Giannis, A., Chen, A., Zhang, J., Chang, V., Wang, J., 2016. Determination of urine-derived odourous compounds in a source separation sanitation system. Journal of Environmental Science. in press.

Liu, J., Schulz, H., Brandl, S., Miehtke, H., Huwe, B., Glaser, B., 2012. Short-term effect of biochar and compost on soil fertility and water status of a Dystric Cambisol in NE Germany under field conditions. Journal of Plant Nutrition and Soil Science 175, 698-707.

López-Cano, I., Roig, A., Cayuela, M.L., Alburquerque, J.A., Sánchez-Monedero, M.A., 2016. Biochar improves N cycling during composting of olive mill wastes and sheep manure. Waste Manage.

Lu, Q., He, Z.L., Stoffella, P.J., 2012. Land application of biosolids in the USA: A review. Applied Environmental and Soil Sciences 2012.

174

Ma, H., Liu, J., Zou, D., Chen, J., Wang, Q., 2010. Application of compound microbial preparations in composting with lactic acid fermentation residue from kitchen waste. Chemical and Biochemical Engineering Quarterly 24, 481-487.

MacLeod, R., Snell, E., 1947. Some mineral requirements of the lactic acid bacteria. Journal of Biological Chemistry 170, 351-365.

Maurer, M., Pronk, W., Larsen, T., 2006. Treatment processes for source-separated urine. Water Research 40, 3151-3166.

Mbah, C., Onweremadu, E., 2009. Effect of organic and mineral fertilizer inputs on soil and maize grain yield in an acid Ultisol in Abakaliki-South Eastern Nigeria. American Eurasia Journal of Agronomy 2, 7-12.

Mee, J., Brooks, C., Stanley, R., 1979. Amino acid and fatty acid composition of cane molasses. J. Sci. Food. Agr. 30, 429-432.

MHRF, 2005. Methods of microbiological control of soils (in Russian).State System for sanitary-epidemiological norms of Russian Federation. Federal Centre of State Epidemiological Supervision, Ministry of Health of Russian Federation, Moskow.

Michetti, P., Dorta, G., Wiesel, P., Brassart, D., Verdu, E., Herranz, M., Felley, C., Porta, N., Rouvet, M., Blum, A., 1999. Effect of whey-based culture supernatant of *Lactobacillus acidophilus* (johnsonii) La1 on *Helicobacter pylori* infection in humans. Digestion 60, 203-209.

Mihelcic, J.R., Fry, L.M., Shaw, R., 2011. Global potential of phosphorus recovery from human urine and feces. Chemosphere 84, 832-839.

Mineev, V., Sychiov, B., Amelyanchik, O., Bolysheva, T., Gomonova, H., Durynina, E., Egorov, B., Egorova, E., Edemskaia, N., Carpova, E., Prijukova, B., 2001. Practice on Agricultural Chemistry (in Russian). Moskow State University, Moskow.

Misselbrook, T., Clarkson, C., Pain, B., 1993. Relationship between concentration and intensity of odours for pig slurry and broiler houses. Journal of Agricultural Engineering Research 55, 163-169.

Mitelut, A.C., Popa, M.E., 2011. Seed germination bioassay for toxicity evaluation of different composting biodegradable materials. Roman Biotechnological Letters. 16, 121-129.

Mnkeni, P., Austin, L., 2009. Fertilizer value of human manure from pilot urine-diversion toilets. Water South Arica 35, 133-138.

Mobley, H., Hausinger, R., 1989. Microbial ureases: significance, regulation, and molecular characterization. Microbiological Reviews 53, 85-108.

Mokoena, M., Mutanda, T., Olaniran, A., 2016. Perspectives on the probiotic potential of lactic acid bacteria from African traditional fermented food and beverages. Food and Nutrition Research 60.

Morella, E., Foster, V., Banerjee, S., 2009. Climbing the ladder: The state of sanitation in sub-Saharan Africa. In: Diagnostic, A.I.C. (Ed.), Background Paper. World Bank, Washington.

Morrison, I., 1988. Influence of some chemical and biological additives on the fibre fraction of lucerne on ensilage in laboratory silos. Journal of Agricultural Sciences 111, 35-39.

Muga, H.E., Mihelcic, J.R., 2008. Sustainability of wastewater treatment technologies. Journal of Environmeantal Management 88, 437-447.

Muller, V., 2001. Bacterial fermentation. Nature Publishing Group, online.

Mulvaney, R., Khan, S., Ellsworth, T., 2009. Synthetic nitrogen fertilizers deplete soil nitrogen: a global dilemma for sustainable cereal production. Journal of Environmental Quality 38, 2295-2314.

Munroe, G., 2007. Manual of on-farm vermi-composting and vermiculture. Dalhousie University, Truro.

Mupondi, L., Mnkeni, P., Muchaonyerwa, P., 2010. Effectiveness of combined thermophilic composting and vermi-composting on biodegradation and sanitization of mixtures of dairy manure and waste paper. African Journal of Biotechnology 9, 4754-4763.

Murphy, P.T., Moore, K.J., Richard, T.L., Bern, C.J., Brumm, T.J., 2007. Use of Swine Manure to Improve Solid-State Fermentation in an Integrated Storage and Conversion System for Corn Stover. Transactions of the ASABE 50, 1901-1906.

Murray, A., Buckley, C., 2010. Designing reuse-oriented sanitation infrastructure: the design for service planning approach. In: Dreschsel, P., Scott, C., Raschid-Sally, L., Redwood, M., Batiri, A. (Eds.), Wastewater Irrigation and Health. Assesing and mitigating risk in low-income countries. Earthscan, London, pp. 303-319.

Murthy, K.N., Malini, M., Savitha, J., Srinivas, C., 2012. Lactic acid bacteria (LAB) as plant growth promoting bacteria (PGPB) for the control of wilt of tomato caused by *Ralstonia solanacearum*. Pest Management in Horticultural Ecosystems 18, 60-65.

Nasiru, A., Ibrahim, M.H., Ismail, N., 2014. Nitrogen losses in ruminant manure management and use of cattle manure vermicast to improve forage quality. International Journal of Recycling of Organic Waste in Agriculture 3, 1-7.

National Bureau of Statistics of the Republic of Moldova, 2016a. Mineral and organic fertilizers used in agricultural enterprises and in farms National Bureau of Statistics of the Republic of Moldova, Chisinau.

National Bureau of Statistics of the Republic of Moldova, 2016b. Generation and use of waste in enterprises and organizations by Type and Movement of waste. National Bureau of Statistics Chisinau.

National Bureau of Statistics of the Republic of Moldova, 2014. Agriculture. Plant production. Yield per hectare of agricultural crops, by categories of producers (quintals) (2006-2013) Chisinau.

Nedealcov, M., Răileanu, V., Chirică, L., Cojocari, R., Mleavaia, G., Sîrbu, R., Gămureac, A., Rusu, V., 2013. Climate resources of the Republic of Moldova. Atlas (in Romanian and English). Stiinta, Chisinau.

Niamsiru, N., Batt, C., 2000. Dairy products. In: Schaecher, M. (Ed.), Encyclopaedia of Microbiology. San Diego University, San Diego.

Niwagaba, C., Kulabako, R.N., Mugala, P., Jonsson, H., 2009 a. Comparing microbial die-off in separately collected faeces with ash and sawdust additives. Waste Management 29, 2214-2219.

Niwagaba, C., Nalubega, M., Vinneras, B., Sundberg, C., Jonsson, H., 2009 b. Bench-scale composting of source-separated human faeces for sanitation. Waste Management 29, 585-589.

Niwagaba, C., Nalubega, M., Vinnerås, B., Sundberg, C., Jönsson, H., 2009 c. Substrate composition and moisture in composting source-separated human faeces and food waste. Environmental Technology 30, 487-497.

Niwagaba, C.B., 2009 d. Treatment technologies for human feaces and urine. Faculty of Natural Resources and Agricultural Sciences Department of Energy and Technology Swedish University of Agricultural Sciences, Uppsala, p. 91.

Noble, R., Jones, P.W., Coventry, E., Roberts, S.R., Martin, M., Alabouvette, C., 2004. Investigations of the Effect of the Composting Process on Particular Plant, Animal and Human Pathogens known to be of Concern for High Quality End-Uses. Oxon, UK, pp. 1-40.

Noike, T., Mizuno, O., 2000. Hydrogen fermentation of organic municipal wastes. Water Science and Technology 42, 155-162.

Nordin, A., 2010. Ammonia sanitisation of human excreta. Energy and Technology. Swedish University of Agricultural Sciences, Uppsala.

Novak, J., Spokas, K., Cantrell, K., Ro, K., Watts, D., Glaz, B., Busscher, W., Hunt, P., 2014. Effects of biochars and hydrochars produced from lignocellulosic and animal manure on fertility of a Mollisol and Entisol. Soil Use Management 30, 175-181.

Novotny, E., Hayes, M., Madari, B., Bonagamba, T., deAzevedo, E., de Souza, A., Song, G., Nogueiraf, C., Mangrich, A., 2009. Lessons from the Terra Preta de Índiosof the Amazon Region for the Utilisation of Charcoal for Soil Amendment. Journal of Brazilian Chemal Society 20, 1003-1010.

Nsamba, H., Hale, S., Cornelissen, G., Bachman, R., 2015. Sustainable Technologies for Small-Scale Biochar Production - A Review. Journal of Sustainable Bioenergy Systems 5, 10-31.

OECD, 2013. Assessing the impact of climate change on water supply sources and WSS systems in Moldova and inventory possible adaptation measures (Task 1). European

Commission (DG Environment) http://ec.europa.eu/environment/marine/international-cooperation/pdf/Moldova_Task%201_Final_EN_26%20Feb.pdf.pdf.

Ong, H., Chew, B., Suhaimi, M., 2001. Effect of effective microorganisms on composting characteristics of chicken manure. Journal of Tropical Agriculture and Food Science 29, 189-196.

Otterpohl, R., Buzie, C., 2011. Wastewater: Reuse-oriented wastewater systems - Low- and high-tech approaches for urban areas. In: Letcher, T., Vallero, D. (Eds.), Waste, a Handbook for Management. Academic Press, Boston, MA, USA, pp. 127–136.

Otterpohl, R., Buzie, C., 2013. Treatment of the solid fraction. In: Larsen, T., Udert, K., Lienert, J. (Eds.), Source Separation and Decentralization for Wastewater Management. IWA Publishing, London, pp. 259-273.

Ottosen, L.D., Poulsen, H.V., Nielsen, D.A., Finster, K., Nielsen, L.P., Revsbech, N.P., 2009. Observations on microbial activity in acidified pig slurry. Biosyst Engineer 102, 291-297.

Pandey, A., Soccol, C.R., Mitchell, D., 2000. New developments in solid state fermentation: I-bioprocesses and products. Proceedings of Biochemistry 35, 1153-1169.

Papajova, I., Juris, P., Zilakova, J., Ilaska, I., 2000. Survival of model helminth Ascaris suum eggs in the ensilaging process of grass biomass under operating conditions. Agroinstitut NItra, Center for Information Services and Technologies 46, 519.

Pape, J., 1970. Plaggen soils in the Netherlands. Geoderma 4, 229-255.

Park, H.Y., DuPonte, M., 2008. How to cultivate microorganisms., Biotechnology. College of Tropical Agriculture and Human Resources, Hawai.

Park, J., Bolan, N., Malavarapu, M., Naidu, R., 2010. Enhancing the solubility of insoluble phosphorus compounds by phosphate solubilizing bacteria. 19th World Congress of Soil Science. Soil solutions for a changing world, Brisbane, Australia, pp. 65-68.

Partha, N., Sivasubramanian, V., 2006. Recovery of chemicals from pressmud - a sugar industry waste. Indian Chemical Engineering 48, 160-163.

Pei-Sheng, Y., Hui-Lian, X., 2002. Influence of EM Bokashi on nodulation, physiological characters and yield of peanut in nature farming fields. Journal of Sustainable Agriculture 19, 105-112.

Pepper, I.L., Brooks, J.P., Gerba, C.P., 2006. Pathogens in biosolids. Advances in Agronomy 90, 1-41.

Perelo, L.W., Munch, J.C., 2005. Microbial immobilisation and turnover of 13 C labeled substrates in two arable soils under field and laboratory conditions. Soil Biology and Biochemistry 37, 2263-2272.

Petersen, S., Andersen, A., Eriksen, J., 2012. Effects of cattle slurry acidification on ammonia and methane evolution during storage. Journal of Environmental Quality 41, 88-94.

Petkova, N., Denev, P., 2012. Evaluation of fructan plants Lactuca serriola L. and Sonchus oleraceus L. Scientific Papers, Series D Animal Science LVI, 125-131.

Pfeiffer, D., 2006. Eating fossil fuel. Oil, food and the coming crisis in agriculture. New Society Publishers, Gabriola Island, Canada.

Pimentel, D., 2006. Soil erosion: a food and environmental threat. Environment, Development and Sustainability 8, 119-137.

Plămădeală, V., Rusu, A., Zubcov, E., Bilețchi, L., Bulat, L., Bîstrova, N., Palamarciuc, N., Șubernețchi, I., Bagrin, N., Ungureanu, G., 2011. Provisional recommendations for application of human urine in agriculture (in Romanian) (Recomandări provizorii privind aplicarea urinei umane în agricultură). In: IPAPS "Nicolae Dimo", EcoTox, Institutul Zoologie, SDC (Eds.). Elena V.I., Chisinau.

Pontin, G., Daly, M., Duggan, C., 2003. Disposal of organic kitchen food waste in the Canterbury/Christchurch region of New Zealand with an EM-Bokashi composting system. Seventh International Conference on Kyusei Nature Farming. Proceedings of the conference Asia Pacific Natural Agriculture Network (APNAN), Christchurch, New Zealand, pp. 275-278.

Popović, M., Grdiša, M., Hrženjak, T.M., 2005. Glycolipoprotein G-90 obtained from the earthworm Eisenia foetida exerts antibacterial activity. Veterinary Arhives 75, 119-128.

Potop, V., 2011. Evolution of drought severity and its impact on corn in the Republic of Moldova. Theoretical and Applied Climatology 105, 469-483.

Prabhu, M., Horvat, M., Lorenz, L., Otterpohl, R., Bettendorf, T., Mutnuri, S., 2014. Effect of terra preta compost on growth of Vigna radiate 1st International Conference on Terra preta sanitation. DBU, Hamburg, Germany.

Pradhan, S.K., Holopainen, J.K., Heinonen-Tanski, H., 2009. Stored human urine supplemented with wood ash as fertilizer in tomato (*Solanum lycopersicum*) cultivation and its impacts on fruit yield and quality. Journal of Agricultural and Food Chemistry 57, 7612-7617.

Pradhan, S.K., Nerg, A.M., Sjöblom, A., Holopainen, J.K., Heinonen-Tanski, H., 2007. Use of Human Urine Fertilizer in Cultivation of Cabbage (*Brassica oleracea*) – Impacts on Chemical, Microbial, and Flavor Quality. Journal of Agricultural and Food Chemistry 55, 8657-8663.

Prakash, M., Karmegam, N., 2010. Vermistabilization of pressmud using *Perionyx ceylanensis* Mich. Bioresource Technology 101, 8464-8468.

Qian, P., Schoenau, J., 2002. Availability of nitrogen in solid manure amendments with different C: N ratios. Canadian Journal of Soil Science 82, 219-225.

Quora, 2016. How many kilograms of compost is the output of 1 ton of compostable waste? The Quora Blog, https://www.quora.com/How-many-kilograms-of-compost-is-the-output-of-1-ton-of-compostable-waste, Quora's mission is to share and grow the world's knowledge.

Ramírez, G., Martínez, R., Herradora, M., Castrejón, F., Galvan, E., 2005. Isolation of Salmonella spp. from liquid and solid excreta prior to and following ensilage in ten swine farms located in central Mexico. Bioresource Technology 96, 587-595.

Rasool, S., Hanjra, S., Jamil, A., 1996. Effect of ensiling sudax fodder with broiler litter and Candida yeast on the changes in different fibre fractions. Animal Feed Science and Technology 57, 325-333.

Rattanachaikunsopon, P., Phumkhachorn, P., 2010. Lactic acid bacteria: their antimicrobial compounds and their uses in food production. Ann Biol Res 1, 218-228.

Raymond, J., Siefert, J.L., Staples, C.R., Blankenship, R.E. 2004. The natural history of nitrogen fixation. Molecular Biology and Evolution, 21(3), 541-554.

Reckin, J., 2010. New insights in matters of plant nutrition, soil microbes and their role in recycling of human excreta and regenerating soil fertility. Sustainable sanitation and water management toolbox.

Redlinger, T., Graham, J., Corella-Barud, V., Avitia, R., 2001. Survival of faecal coliforms in dry-composting toilets. Applied Environmental Microbiology 67, 4036-4040.

Richard, T., Walker, L., Gossett, J., 2006. Effects of oxygen on aerobic solid-state biodegradation kinetics. Biotechnology Progress 22, 60-69.

Rieck, C., von Münch, E., Hoffman, H., 2012. Technology review of Urine-diverting Dry Toilets (UDDTs). Deutsche Gesellschaft für Internationale Zusammenarbeit (GIZ) GmbH, Eschborn.

Rijsberman, F., 2006. Water scarcity: fact or fiction? Agricul Water Manage 80, 5-22.

Rillig, M.C., Wagner, M., Salem, M., Antunes, P.M., George, C., Ramke, H.-G., Titirici, M.-M., Antonietti, M., 2010. Material derived from hydrothermal carbonization: effects on plant growth and arbuscular mycorrhiza. Applied Soil Ecology 45, 238-242.

Rodushkin, I., Ödman, F., 2001. Application of inductively coupled plasma sector field mass spectrometry for elemental analysis of urine. Journal of Trace Elements and Medical Biology 14, 241-247.

Ronsse, F., Van Hecke, S., Dickinson, D., Prins, W., 2013. Production and characterization of slow pyrolysis biochar: influence of feedstock type and pyrolysis conditions. GCB Bioenergy 5, 104-115.

Rose, C., Parker, A., Jefferson, B., Cartmell, E., 2015. The characterization of feces and urine: A review of the literature to inform advanced treatment technology. Crital Reviews in Environmental Science and Technology 45, 1827-1879.

Rosemarin, A., 2010. Peak phosphorus and eutrophication of surface waters: A symptom of disconnected policies to govern agricultural and sanitation practices.

Roy, R.N., Misra, R.V., Montanez, A., 2002. Decreasing reliance on mineral nitrogen-yet more food. Ambio 31, 177-183.

182

Runge, E., 1968. Effects of rainfall and temperature interactions during the growing season on corn yield. Agronomy Journal 60, 503-507.

Rusco, E., Jones R, G, B., 2001. Organic matter in the soils of Europe: present status and future trends. Joint Research Centre European Commission, Ispra, Italy.

Saeed, H., Salam, I., 2013. Current limitations and challenges with lactic acid bacteria: a review. Food Nutrition Science 4, 73-87.

Saman, P., Fuciños, P., Vázquez, J.A., Pandiella, S.S., 2011. Fermentability of brown rice and rice bran for growth of human *Lactobacillus plantarum* NCIMB 8826. Food Technology and Biotechnology 49, 128.

Samer, M., Mostafa, E., Hassan, A., 2014. Slurry treatment with food industry wastes for reducing methane, nitrous oxide and ammonia emissions. Misr Journal of Agricultural Engineering 31, 1523-1548.

Sancom, SA Sugar beet molasses in big quantities (in Romanian). Melasa sfecla de zahar. Available from https://www.bizoo.ro/firma/trinex/vanzare/1357217/melasa-din-sfecla-de-zahar-in-cantitati-mari. Last visit on 03 February, 2017.

Saranraj, P., 2014. Lactic acid bacteria and its antimicrobial properties a review. Internatioanl Journal of Pharmaceutical Biology Archiv. 4, Available Online at www.ijpba.info.

Sato, T., Manzoor Q., Ymamoto, S., Endo, T., Zahoor, A., 2013. Global, regional and country level need for data on wastewater generation, treatment and use. Agriculture, Water Management 130, 1-13.

Savijoki, K., Ingmer, H., Varmanen, P., 2006. Proteolytic systems of lactic acid bacteria. Applied Microbiology and Biotechnology 71, 394-406.

Scerbacov, A., Vassenev, I., 2000. Anthropogenic evolution of chernoziom soils (in Russian). Voronej State Univeristy, Voronej, Russia.

Scheinemann, H., Dittmar, K., Erfurt, K., Stoeckel, F., Krueger, M., 2013. Hygienization and nutrient conservation of sewage sludge of cattle manure by fermentation. 1st International Conference on Terra preta sanitation, Hamburg, p. 51.

Scheinemann, H., Dittmar, K., Stöckel, F., Müller, H., Krüger, M., 2015. Hygienisation and nutrient conservation of sewage sludge or cattle manure by lactic acid fermentation. PLoS One 10, doi:10.1371/journal.pone.0118230.

Sherlock, R.R. 1984. Dynamics of ammonia volatilization and nitrous oxide production from urine patches in grazed pastures, Lincoln College, University of Canterbury. Canterbury, UK.

Schertenleib, R., 2005. From conventional to advanced environmental sanitation. Water Science & Technology 51, 7-14.

Schmidt, H. P., Kammann, C., Niggli, C., Evangelou, M.W., Mackie, K.A., Abiven, S., 2014. Biochar and biochar-compost as soil amendments to a vineyard soil: Influences on plant growth, nutrient uptake, plant health and grape quality. Agriculture, Ecosystems & Environment 191, 117-123.

Schmidt, H., Pandit, B., Martinsen, V., Cornelissen, G., Conte, P., Kammann, C., 2015. Fourfold increase in pumpkin yield in response to low-dosage root zone application of urine-enhanced biochar to a fertile tropical soil. Agriculture 5, 723-741.

Schneider, J., Kaltwasser, H., 1984. Urease from *Arthrobacter oxydans*, a nickel-containing enzyme. Archives of Microbiology 139, 355-360.

Schönning, C., Stenström, T.A., 2004. Guidelines for the safe use of urine and faeces in ecological sanitation systems. EcoSanRes Programme.

Schroeder, J., 2004. Silage fermentation and preservation. NDSU Extension Service Fargo, North Dakota.

Schuetze, T., Santiago-Fandino, 2014. Terra preta sanitation: a key component for sustainability in the urban environment. Sustainability 6, 7725-7750.

Schwarz, W., 2001. The cellulosome and cellulose degradation by anaerobic bacteria. Applied microbiology and biotechnology 56, 634-649.

Servin, A., 2004. Antagonistic activities of lactobacilli and bifidobacteria against microbial pathogens. FEMS Microbiological Reviews 28, 405-440.

Sghir, A., Gramet, G., Suau, A., Rochet, V., Pochart, P., Dore, J., 2000. Quantification of bacterial groups within human faecal flora by oligonucleotide probe hybridization. Applied Environmental Microbiology 66, 2263-2266.

Sgouras, D., Maragkoudakis, P., Petraki, K., Martinez-Gonzalez, B., Eriotou, E., Michopoulos, S., Kalantzopoulos, G., Tsakalidou, E., Mentis, A., 2004. In vitro and in vivo inhibition of *Helicobacter pylori* by *Lactobacillus casei* strain Shirota. Applied Environmental Microbiology 70, 518-526.

Sherlock, R.R., 1984. Dynamics of ammonia volatilization and nitrous oxide production from urine patches in grazed pastures. Lincoln College, University of Canterbury, Canterbury, Uk.

Sijpesteijn, A., 1951. On *Ruminococcus flavefaciens*, a cellulose-decomposing: bacterium from the rumen of sheep and cattle. Microbiology 5, 869-879.

Singh, B., Cowie, A., 2014. Long-term influence of biochar on native organic carbon mineralisation in a low-carbon clayey soil. Scientific Reports 4, 3687-3700.

Siqueira, P., Karp, S., Carvalho, J., Sturm, W., Rodríguez-León, J., Tholozan, J., Singhania, R., Pandey, A., Soccol, C., 2008. Production of bio-ethanol from soybean molasses by *Saccharomyces cerevisiae* at laboratory, pilot and industrial scales. Bioresource Technology 99, 8156-8163.

Smetanova, A., Dotterweich, M., Diehl, D., Ulrich, U., Fohrer, N., 2012. Influence of biochar and terra preta substrates on wettability and erodibility of soils. EGU General Assembly. Geophysical Research Abstracts.

Snyman, L., du Preez, R., Calitz, I., 1986. Fermentation characteristics and chemical composition of silage made from different ratios of cattle manure and maize residues. South African Journal of Animal Science 16, 83-86.

Sokolov, A.V., Askinazi, P.L., 1965. Agrochemical methods of soil investigation (in Russian). Science Inc., Moskow.

Solaimalai, A., Baskar, M., Ramesh, P., Ravisankar, N., 2001. Utilisation of press mud as soil amendment and organic manure – a review. Agricultural Reviews 22, 25-32.

Sombroek, W., Kern, D., Rodrigues, T., Cravo, M.S., Jarbas, T.C., Woods, W., Glaser, B., 2002. Terra Preta and Terra Mulata: pre-Columbian Amazon kitchen middens and agricultural fields, their sustainability and their replication. 17th WCSS, Thailand.

Somers, E., Amke, A., Croonenborghs, A., Overbeek, L., Vaderleyden, J., 2007. Lactic acid bacteria in organic agricultural soils. Rhizosphere.

Southgate, D., Durnin, J., 1970. Calorie conversion factors. an experimental reassessment of the factors used in the calculation of the energy value of human diets. British Journal of Nutrition 24, 517-535.

Sokolov, A.V., Askinazi, P.L., 1965. Agrochemical methods of soil investigation (in Russian). Science Inc., Moskow.

Spiertz, J.H., De Vos, N.M., 1983. Agronomical and physiological aspects of the role of nitrogen in yield formation of cereals. Plant and Soil 75, 379-391.

Spokas, K.A., Novak, J.M., Venterea, R.T., 2012. Biochar's role as an alternative N-fertilizer: ammonia capture. Plant and Soil 350, 35-42.

Spoor, M., Izman, F., 2009. Land reform and interlocking agricultural markets in Moldova. In: Spoor, M. (Ed.), The political economy of rural livelihoods in transition economies. Taylor and Francis Group, London and New York.

Stadnic, S., 2010. Pedology with the basis of geology (in Romanian). State University Alecu Russo, Balti.

Stadnik, V., 2010. Pedogeography with Pedology Basis (in Romanian). Aleco Russo University, Balti, Moldova.

Stein, W., Carey, G., 1953. A chromatographic investigation of the amino acid constituents of normal urine. J. Biol. Chem. 201, 45-58.

Steiner, C., Das, K., Melear, N., Lakly, D., 2010. Reducing nitrogen loss during poultry litter composting using biochar. Journal of Environmental Quality 39, 1236-1242.

Stenmarck, A., Jensen, C., Quested, T., Moates, G., 2016. Colophon. Estimates of the European food waste level. Fusions, Stockholm.

Stevenson, F.J., 1994. Humus chemistry: genesis, composition, reactions. John Wiley & Sons, Dublin.

Stoeckl, M., Roggentin, P., Bettendorf, T., Otterpohl, R., 2013. Assessment of hygienisation of faecal matter during terra preta inspired vermi-composting by qualitative identification of Salmonellaspec. . In: Bettendorf, T., Wendland, C., Otterpohl, R. (Ed.), 1st International Terra Preta Sanitation Conference. GFEU and WECF Hamburg.

Strauss, M., Drescher, S., Zurbrügg, C., Montangero, A., Cofie, O., Drechsel, P., 2003. Co-composting of faecal sludge and municipal organic waste: a literature and state-of-knowledge review. Swiss Federal Institute of Environmental Science and Technology (EAWAG), Duebendorf.

Strong, K., Osicka, T., Comper, W., 2005. Urinary-peptide excretion by patients with and volunteers without diabetes. Journal of Laboratory and Clinical Medicine 145, 239-246.

Syers, K., Bekunda, M., Cordell, D., Corman, J., Johnston, J., Rosemarin, A., Salcedo, I., 2011. Phosphorus and food production. UNEP Year Book, 34-45.

Tabak, S., Maghnia, D., Bensoltane, A., 2012. The Antagonistic activity of the lactic acid bacteria (*Streptococcus thermophilus*, *Bifidobacterium bifidum* and *Lactobacillus bulgaricus*) against *Helicobacter pylori* responsible for the gastroduodenals diseases. Journal of Agricultural Science and Technology A2, 709-715.

Taghizadeh-Toosi, A., Clough, T., Sherlock, R., Condron, L., 2011. Biochar adsorbed ammonia is bioavailable. Plant and Soil 350, 1-13.

Tan, Z., Lal, R., Wiebe, K., 2005. Global soil nutrient depletion and yield reduction. Journal of Sustainable Agriculture 26, 123-146.

Tchobanoglous, G., Burton, S., Stensel, H.D., 2003. Wastewater engineering: treatment and reuse. New York.

The World Bank, 2014. Moldova national water supply and sanitation project. World Bank, http://documents.worldbank.org/curated/en/2014/05/19883941/moldova-national-water-supply-sanitation-project.

Tilley, E., Ulrich, L., Lüthi, C., Reymond, P., Zurbrügg, C., 2014. Compendium of sanitation systems and technologies. Eawag.

Tim, A., 2015. Managing food waste for sustainability: landfill versus composting. DS Food Systems, Sustainability and Climate Change. Tim, A., UW Madison Campus.

Tiqua, S., Tam, N., Hodgkiss, I., 1996. Effects of composting on phytotoxicity of spent pig-manure sawdust and litter. Environmental Pollution 93, 249-256.

Tognetti, C., Laos, F., Mazzarino, M., Hernandez, M., 2005. Composting vs. vermicomposting: a comparison of end product quality. Compost Science & Utilization 13, 6-13.

Topoliantz, S., Ponge, J.-F., Ballof, S., 2005. Manioc peel and charcoal: a potential organic amendment for sustainable soil fertility in the tropics. Biology and Fertility of Soils 41, 15-21.

Triastuti, J., Sintawardani, N., Irie, M., 2009. Characteristics of composted bio-toilet residue and its potential use as a soil conditioner. Indones Journal of Agricultural Science 10, 73-79.

Troccaz, M., Niclass, Y., Anziani, P., Starkenmann, C., 2013. The influence of thermal reaction and microbial transformation on the odour of human urine. Flavour Fragrance Journal 28, 200-211.

Turral, H., Burke, J., Faures, J., 2011. Climate change, water and food security. FAO Water Report FAO, Rome.

Udawatta, R.P., Motavalli, P.P., Garrett, H.E., Krstansky, J.J., 2006. Nitrogen losses in runoff from three adjacent agricultural watersheds with claypan soils. Agriculture, Ecosystems & Environment 117, 39-48.

Udert, K., Wächter, M., 2012. Complete nutrient recovery from source-separated urine by nitrification and distillation. Water Research 46, 453-464.

Udert, K., Larsen, T., Gujer, W., 2006. Fate of major compounds in source-separated urine. Water Science & Technology 54, 413-420.

Udert, K.M., Larsen, T.A., Biebow, M., Gujer, W., 2003. Urea hydrolysis and precipitation dynamics in a urine-collecting system. Water Research 37, 2571-2582.

UNECE, SDC, 2011 Setting targets and target dates under the Protocol on Water and Health in the Republic of Moldova, Eco-Tiras, Chisinau.

Ursu, A., P., V., Curcubăt, S., 2009. The pedologic potential of the North districts of silvostepe (in Romanian). Buletinul Academiei de Stiinte a Moldovei 1, 147-152.

van der Poel, P.W., Schiweck, H., Schwartz, T., 1998. Sugar technology: Beet and cane sugar manufacture. Verlag, Berlin.

Van Laere, A., Ende, V.D., 2002. Inulin methabolism in dicots: cicory as a model system. Plant, Cell and Environment 25, 803-881.

Verheijen, F., Jeffery, S., Bastos, A., Van der Velde, M., Diafas, I., 2010. Biochar application to soils: a critical scientific review of effects on soil properties, processes and functions. Joint Research Centre, Ispra, Italy.

Verigo, C.A., Razumova, L.M., 1973. Soil moisture. Applicable for rural collective farms. Status: good (in Russian). Gydrometeoizdat, Leningrad.

Vinneras, B., Bjorklund, A., Jonsson, H., 2003. Thermal composting of faecal matter as treatment and possible disinfection method-laboratory-scale and pilot-scale studies. Bioresource Technology 88, 47-54.

Virto, I., Imaz, M., Fernández-Ugalde, O., Gartzia-Bengoetxea, N., Enrique, A., Bescansa, P., 2014. Soil degradation and soil quality in Western Europe: current situation and future perspectives. Sustainability 7, 313-365.

Visser, R., Holzapfel, W., Bezuidenhout, J., Kotzé, J., 1986. Antagonism of lactic acid bacteria against phytopathogenic bacteria. Appled Environmental Microbiology 52, 552-555.

Vronskih, M., 2014. Climate change and its impact on soil productivity in Moldova. In: Dent, D. (Ed.), Soil as a world heritage. Springer, Dordrecht, pp. 101-117.

Wachendorf, C., Taube, F., Wachendorf, M., 2005. Nitrogen leaching from ^{15}N labeled cow urine and dung applied to grassland on a sandy soil. Nutrient Cycling in Agroecosysts 73, 89-100.

Wagenstaller, M., Buettner, A., 2013. Quantitative determination of common urinary odorants and their glucuronide conjugates in human urine. Metabolites 3, 637-657.

Wang, C., Lu, H., Dong, D., Deng, H., Strong, P., Wang, H., Wu, W., 2013. Insight into the effects of biochar on manure composting: evidence supporting the relationship between N_2O emission and denitrifying community. Environmental Science and Technology 47, 7341-7349.

Wang, W., Wang, X., Liu, J., Ishii, M., Igarashi, Y., Cui, Z., 2007. Effect of oxygen concentration on the composting process and maturity. Compost Science & Utilization 15, 184-190.

Wang, Q., Yamabe, K., Narita, J., Morishita, M., Ohsumi, Y., Kusano, K., Shirai, Y., Ogawa, H., 2001. Suppression of growth of putrefactive and food poisoning bacteria by lactic acid fermentation of kitchen waste. Process Biochemistry 37, 351-357.

Watkinson, A., Murby, E., Costanzo, S., 2007. Removal of antibiotics in conventional and advanced wastewater treatment: implications for environmental discharge and wastewater recycling. Water Research 41, 4164-4176.

Wendland, C., Deegener, S., Jorritsma, F., 2011. Experiences with urine diverting dry toilets (UDDTs) for households, schools and kindergarten in Eastern Europe, the Caucasus and Central Asia (EECCA). Sustainable Sanitation Practices 6, 16-22.

Wernli, M., 2014. Growing lactobacilli inocula. Self-suficient fermentation of organic matter in terra preta sanitation. Canberra University, Bruce, Canberra, pp. 1-10.

Westermann, D., Crothers, S., 1980. Measuring soil nitrogen mineralization under field conditions. Agronomy Journal 72, 1009-1012.

White, A., Hess, W., 1956. Paper chromatographic detection of sugars in normal and dystrophic human urines. Archieves Biochemistry and Biophysics 64, 57-66.

WHO, 2006. Guidelines for the safe use of wastewater, excreta and greywater. World Health Organization, Geneva.

Wiedner, K., Glaser, B., 2015a. Traditional use of biochar. In: Lehmann, J., Joseph, S. (Eds.), Biochar for Environmental Management: Science, Technology and Implementation. Earthscan, New York, p. 15.

Wiedner, K., Schneerweis, J., Dippold, M., Glaser, B., 2015b. Anthropogenic Dark Earth in Northern Germany - The Nordic Analogue to terra preta de Índio in Amazonia. Catena 132, 114-125.

Williams, T., Akse, J., Handley, N., Glock, D., 2011. Nonhazardous urine pretreatment method for future exploration. 41 International Conference on Environmental Systems, Portland, Oregon.

Winblad, U., 1997. Towards an ecological approach to sanitation. Swedish international development cooperation agency (Sida), Stockholm.

Winblad, U., Simpson-Hebert, M., 2004. Ecological sanitation - revised and enlarged edition. Stockholm Institute of Environment, Stockholm, Sweden.

Windberg, C., Yemaneh, A., Otterpohl, R., 2013. Terra preta sanitation: a new tool for sustainable sanitation in urban areas. 36th WEDC International Conference Nakuru, Kenya.

Winker, M., Vinnerås, B., Muskolus, A., Arnold, U., Clemens, J., 2009. Fertilizer products from new sanitation systems: Their potential values and risks. Bioresource Technology 100, 4090-4096.

Wolf, R., 2013. Application of fermented urine to build up terra preta humus in a permaculture park and social impact on the communities involved. 1st International Conference on Terra preta sanitation. TUHH, Hamburg, pp. 35-36.

Woods, W., 2003. History of anthrosol research. In: Lehmann, J., Kern, D., Glaser, B., Woods, W. (Eds.), Amazonian dark earths: origin, properties, management. Kluwer Academic Publishers, Dordrecht pp. 3-14.

Xavier, S., Lonsane, B., 1994. Sugar-cane pressmud as a novel and inexpensive substrate for production of lactic acid in a solid-state fermentation system. Applied Microbiological Biotechnology 41, 291-295.

Xiaohou, S., Min, T., Ping, J., Weiling, C., 2008. Effect of EM Bokashi application on control of secondary soil salinization. Water Science and Engineering 1, 99-106.

Yadav, A., Garg, V., 2011. Recycling of organic wastes by employing *Eisenia fetida*. Bioresource Technology 102, 2874-2880.

Yadav, K., Tare, V., Ahammed, M., 2010. Vermi-composting of source-separated human faeces for nutrient recycling. Waste Management 30, 50-56.

Yang, S., Ji, K., Baik, Y., Kwak, W., McCaskey, T., 2006. Lactic acid fermentation of food waste for swine feed. Bioresource Technology 97, 1858-1864.

Yao, Y., Gao, B., Inyang, M., Zimmerman, A., Cao, X.Y., Pullammanappallil, P., Yang, L., 2011. Removal of phosphate from aqueous solution by biochar derived from anaerobically digested sugar beet tailings. Journal of Hazardous Materials 190, 501-507.

Yemaneh, A., 2015. Evaluation of lactic acid fermentation process in terra preta sanitation system and application in Arba Minch, Ethiopia. Hamburg University. Hamburg Technical University, Hamburg.

Yemaneh, A., Bulbo, M., Schmale, C., Otterpohl, R., 2014. Investigation of low-cost sugar supplement for lacti acid fermentation in terra preta sanitation system. In: Bettendorf, T., Wendland, C., Otterpohl, R. (Eds.), 1st International Conference on Terra preta sanitation. DBU, Hamburg, Germany.

Yemaneh, A., Bulbo, M., Schmale, C., Otterpohl, R., 2013. Investigation of low-cost sugar supplements for lactic-acid fermentation. 1st International Conference on Terra preta sanitation, Hamburg, pp. 53-55.

Yemaneh, A., Bulbo, A.M., Factura, H., Buzie, C., Otterpohl, R., 2012. Development of System for Waterless Collection system of Human Excreta by Application of Lactic Acid Fermentation Process in Terra Preta Sanitation 4th Dry Toilet Conference. Global Dry Toilet Association of Finland, Tampere, Finland.

Zavala, M., Funamizu, N., Takakuwa, T., 2002. Characterization of feces for describing the aerobic biodegradation of feces. Journal of Environmental Systems and Engineering 720, 99-105.

Zavala, M., Funamizu, N., Takakuwa, T., 2005. Biological activity in the composting reactor of the bio-toilet system. Bioresource Technology 96, 805-812.

Zhang, H., Tan, S., Wong, W., Ng, C., Teo, C., Ge, L., Chen, X., Yong, J., 2014a. Mass spectrometric evidence for the occurrence of plant growth promoting cytokinins in vermicompost tea. Biological Fertility of Soils 50, 401-403.

192

Zhang, Y.H., Xu, D., Liu, J.Q., Zhao, X.-H., 2014b. Enhanced degradation of five organophosphorus pesticides in skimmed milk by lactic acid bacteria and its potential relationship with phosphatase production. Food Chemistry 164, 173-178.

Zhang, J., Giannis, A., Chang, V., Ng, B., Wang, J., 2013. Adaptation of urine source separation in tropical cities: process optimization and odour mitigation. Journal of Air and Waste Management Association 63, 472-481.

Zhang, A., Cui, L., Pan, G., Li, L., Hussain, Q., Zhang, X., Zheng, J., Crowley, D., 2010. Effect of biochar amendment on yield and methane and nitrous oxide emissions from a rice paddy from Tai Lake plain, China. Agriculture Ecosystems and Environment 139, 469-475.

Zimmer, D., Hofwegen, P., 2006. Costing MDG Target 10 on water supply and sanitation: comparative analysis, obstacles and recommendations. Marceille, France.

Zimmerman, A., Gao, B., Ahn, M., 2011. Positive and negative carbon mineralization priming effects among a variety of biocharamended soils. Soil Biology and Biochemistry 43, 1169-1179.

Zlotnikov, K., Zlotnikov, A., Kaparullina, E., Doronina, N., 2013. Phylogenetic position and phosphate solubilization activity of lactic acid bacteria associated with different plants. Microbiology 82, 393-396.

Zygmunt, B., Bannel, A., 2008. Formation, occurrence and determination of volatile fatty acids in environmental and related samples. 3rd WSEAS Int. Conf. on Waste Management, water pollution, air pollution and indoor climate Corfu Island, Greece.

Appendixes

Appendix A

Corn growth rate as influenced by lacto-fermented mix and biochar in comparison to other fertilizers applied

Dunnett Simultaneous Tests for Level Mean - Control Mean

1st production period

Plant height, cm

Difference of levels	Difference of means	SE of difference	95 % CI	T-value	P-value
1-3	-29.57	1..20	(-33.06; -26.07)	-21.63	0.000
2-3	-14.37	1.20	(-17.86; -10.87)	-11.97	0.000
4-3	-8.10	1.20	(-11.60; -4.60)	-6.75	0.000
5-3	-12.37	1.20	(-15.86; -8.87)	-10.30	0.000
6-3	-17.17	1.20	(-20.66; -13.67)	-14.30	0.000
7-3	-17.90	1.20	(-21.40;-14.40)	-14.91	0.000

Stem diameter, cm

Difference of levels	Difference of means	SE of difference	95 % CI	T-value	P-value
1-3	-677	0.155	(-1.129; -0.224)	-4.35	0.003
2-3	-0.133	0.155	(-0.586; -0.319)	-0.86	0.892
4-3	-0.257	0.155	(-709; 0.196)	-1.65	0.407
5-3	-0.357	0.155	(-0.809; 0.096)	-2.30	0.150
6-3	-0.410	0.155	(-0.863; 0.043)	-2.64	0.082
7-3	-0.443	0.155	(-0.896; 0.009)	-2.85	0.056

Leaf length. cm

Difference of levels	Difference of means	SE of difference	95 % CI	T-value	P-value
1-3	-5.90	1.23	(-9.48; -2.32)	-4.80	0.001
2-3	-3.00	1.23	(-6.58; 0.58)	-2.44	0.117
4-3	-0.34	1.23	(-3.23; 3.92)	0.28	1.000
5-3	-1.83	1.23	(-5.41; 1.74)	-1.49	0.501
6-3	-2.77	1.23	(-0.863; 0.043)	-2.25	0.162
7-3	-5.77	1.23	(-9.34; -2.19)	-4.69	0.002

Leaf width. cm

Difference of levels	Difference of means	SE of difference	95 % CI	T-value	P-value
1-3	-0.467	0.619	(-2.271; 1,337)	-0.75	0.934
2-3	-0.467	0.619	(-2.271; 1,337)	-0.75	0.934
4-3	-0.697	0.619	(-2.501; 1.107)	-1.12	0.741
5-3	-0.600	0.619	(-2.404; 1.204)	-0.97	0.835
6-3	-1.167	0.619	(-2.971; 0.637)	-1.88	0.292
7-3	-1.800	0.619	(-3.604; 0.004)	-2.91	0.051

2nd production period

Plant height, cm

Difference of levels	Difference of means	SE of difference	95 % CI	T-value	P-value
1-3	-9.90	1.24	(-13.54; -6.26)	-7.96	0.000
2-3	2.23	1.24	(-1.40; 5.87)	1.80	0.356
4-3	-1.77	1.24	(-5.40; 1.87)	-1.12	0.582
5-3	4.77	1.24	(1.13; 8.40)	-1.42	0.008
6-3	-7.17	1.24	(-10.80; -3.53)	-5.76	0.000
7-3	1.27	1.24	(-2.37; 4.90)	1.02	0.841
8-3	-3.83	1.24	(-7.47; -0.20)	-3.08	0.037

Stem diameter, cm

Difference of levels	Difference of means	SE of difference	95 % CI	T-value	P-value
1-3	-233	0.202	(-0.823; 0.356)	-1.16	0.757
2-3	0.200	0.202	(-0.389; 0.789)	0.99	0.855
4-3	0.333	0.202	(-0.256; 0.923)	1.65	0.406
5-3	0.343	0.202	(-0.246; 0.933)	1.70	0.008
6-3	0.067	0.202	(-0. 523; 0.656)	0.33	1.000
7-3	0.100	0.202	(-0.489; 0.689)	0.50	0.995
8-3	-0.200	0.202	(-0.789; 0.389)	-0.99	0.855

Leaf length, cm

Difference of levels	Difference of means	SE of difference	95 % CI	T-value	P-value
1-3	-7.07	1.58	(-11.69; -2.44)	-4.47	0.002
2-3	2.47	1.58	(-2.16; 7.09)	1.56	0.493
4-3	-2.87	1.58	(-7.49; 1.76)	-1.81	0.348
5-3	9.07	1.58	(4.44; 13.69)	5.73	0.000
6-3	5.30	1.58	(-0.67; 9.93)	3.35	0.022
7-3	2.67	1.58	(-1.96; 7.29)	1.69	0.417
8-3	-5.83	1.58	(-10.46; -1.21)	-3.69	0.011

Leaf width, cm

Difference of levels	Difference of means	SE of difference	95 % CI	T-value	P-value
1-3	-0.593	0.250	(-1.324; 0.137)	-2.37	0139
2-3	-0.567	0.250	(-1.297; 0.164)	-2.27	0.167
4-3	-400	0.250	(-1.131; 0.331)	-1.60	0.467
5-3	-0.067	0.250	(-0.797; 0.664)	-0.27	1.000
6-3	-0.100	0.250	(-0.831; 0.631)	-0.40	0.999
7-3	-0.667	0.250	(-1.397; 0.064)	-2.67	0.081
8-3	-0.933	0.250	(-1.664; -0.203)	-3.73	0.010

1- Control
2- Lacto-fermented mix
3- Lacto-fermented mix and biochar
4- Cattle manure
5- Mineral fertilizer
6- Stored faeces
7- Stored urine
8- Vermi-composted lacto-fermented mix and biochar

Appendix B

The yield of corn as influenced by lacto-fermented mix and biochar in comparison to other fertilizers.

2013

Difference of levels	Difference of means	SE of difference	95 % CI	T-value	P-value
1-3	-789.7	71.8	(-998.9; -580.4)	-10.99	0.000
2-3	-16	71.8	(-225.3; 193.3)	-0.22	1.000
4-3	-269.3	71.8	(-478.6; -60.1)	-3.75	0.010
5-3	-811.3	71.8	(-1020.6; -602.1)	-11.29	0.000
6-3	-357.3	71.8	(-566.6; -148.1)	-4.97	0.001
7-3	-352.3	71.8	(--561.6; -143.1)	-4.90	0.001

2014

Difference of levels	Difference of means	SE of difference	95 % CI	T-value	P-value
1-3	-1159.3	64.7	(-1348.4; -970.3)	-17.93	0.000
2-3	-693.0	64.7	(-882.1; -503.9)	-10.72	0.000
4-3	-858.3	64.7	(-1047.4; -669.3)	-13.27	0.000
5-3	109.3	64.7	(-79.7; 298.4)	1.69	0.000
6-3	-1274.0	64.7	(-1463.1; -1048.9)	-19.70	0.413
7-3	-648	64.7	(-837.1; -458.9)	-10.02	0.000
8-3	-878.7	64.7	(-1067.7; -689.6)	-13.59	0.000

Appendix C

Soil humus content as influenced by the application of lactofermented mix and biochar in comparison to other fertilizers

Humus content, 0-20 cm, 2013

Dunnett Simultaneous Tests for Level Mean - Control Mean

Difference of levels	Difference of means	SE of difference	95 % CI	T-value	P-value
1-3	-0.113	0.157	(-0.570; 0.343)	-0.72	0.945
2-3	0.177	0.157	(-280; 0.633)	1.13	0.740
4-3	0.013	0.157	(-0.443; 0.470)	0.09	1.000
5-3	-0.213	0.157	(-0.670; 0.243)	-1.36	0.586
6-3	-0.047	0.157	(-0.503; -0.410)	-0.30	0.999
7-3	-0.233	0.157	(-0.690; 0.223)	-1.49	0.504

Humus content, 20-40 cm, 2013

Difference of levels	Difference of means	SE of difference	95 % CI	T-value	P-value
1-3	-0.0.017	0.128	(-0.390; 0.357)	-0.13	1.000
2-3	-0.007	0.128	(-380; 0.367)	-0.05	1.000
4-3	0,233	0.128	(-0.140; 0.607)	1.82	0.320
5-3	-0.350	0.128	(-0.723; 0.023)	-2.73	0.070
6-3	-0.023	0.128	(-0.397; 0.350)	-0.18	1.000
7-3	-0.113	0.128	(-0.487; 0.260)	-0.88	0.879

Humus content, 0-20 cm, 2014

Difference of levels	Difference of means	SE of difference	95 % CI	T-value	P-value
1-3	-0.147	0.181	(-0.677; 0.383)	-0.81	0.937
2-3	0.073	0.181	(-457; 0.603)	0.40	0.998
4-3	0.170	0.181	(-0.360; 0.700)	0.94	0.883
5-3	-0.227	0.181	(-0.757; 0.303)	-1.25	0.697
6-3	0.010	0.181	(-0.520; 0.540)	0.06	1.000
7-3	-0.207	0.181	(-0.737; 0.323)	-1.14	0.769
8-3	-0.040	0.181	(-0.570; 0.490)	-0.22	1.000

Humus content, 20-40 cm, 2014

Difference of levels	Difference of means	SE of difference	95 % CI	T-value	P-value
1-3	-0.147	0.191	(-0.705; 0.411)	-0.77	0.950
2-3	0.073	0.191	(-485; 0.631)	0.38	0.999
4-3	0.170	0.191	(-0.388; 0.728)	0.89	0.905
5-3	-0.227	0.191	(-0.785; 0.331)	-1.19	0.738
6-3	0.010	0.191	(-0.548; 0.568)	0.05	1.000
7-3	-0.207	0.191	(-0.735; 0.351)	-1.08	0.804
8-3	-0.090	0.191	(-0.468; 0.648)	0.47	0.996

Appendix D

Phosphorus content, mg/kg 0-20 cm, 1st production period, 2013

Difference of levels	Difference of means	SE of difference	95 % CI	T-value	P-value
1-3	-10.70	5.24	(-25.96; 4.56)	-2.04	0.228
2-3	-7.60	5.24	(-22.86; 7.66)	-1.45	0.528
4-3	-2.83	5.24	(-18.10; 12.43)	-0.54	0.985
5-3	-1.87	5.24	(-17.13; 13.40)	-0.36	0.998
6-3	-5.00	5.24	(-20.26; 10.26)	-0.95	0.843
7-3	-2.87	5.24	(-18.13; 12.40)	-0.55	0.984

Phosphorus content, mg/kg 20-40 cm, 1st production period, 2013

Difference of levels	Difference of means	SE of difference	95 % CI	T-value	P-value
1-3	1.93	5.68	(-14.61; 18.48)	0.34	0.999
2-3	-0.07	5.68	(-16.61; 16.48)	-0.01	1.000
4-3	1.27	5.68	(-15.28; 17.81)	0.22	1.000
5-3	4.10	5.68	(-12.44; 20.64)	0.72	0.945
6-3	-4.33	5.68	(-20.88; 12.21)	-0.76	0.931
7-3	0.83	5.68	(-15.71; 17.38)	0.15	1.000

Phosphorus content, mg/kg 0-20 cm, 2ndproduction period, 2014

Difference of levels	Difference of means	SE of difference	95 % CI	T-value	P-value
1-3	-31.07	4.65	(-44.66; -17.47)	-6.68	0.000
2-3	-4.40	4.65	(-17.99; 9.19)	-0.95	0.879
4-3	-20.17	4.65	(-33.76; -6.57)	-4.34	0.003
5-3	5.37	4.65	(-8.13; 19.06)	1.18	0.746
6-3	7.37	4.65	(-6.23; 20.96)	1.58	0.477
7-3	2.07	4.65	(-11.53; 15.66)	0.44	0.997
8-3	-15.17	4.65	(-28.76; -1.57)	-3.26	0.026

Phosphorus content, mg/kg 20-40 cm, 2ndproduction period, 2014

Difference of levels	Difference of means	SE of difference	95 % CI	T-value	P-value
1-3	-16.60	3.10	(-26.57; -7.58)	-5.35	0.000
2-3	0.30	3.10	(-8.77; 9.37)	0.10	1.000
4-3	-8.43	3.10	(-17.50; 0.63)	-2.72	0.074
5-3	-5.57	3.10	(-14.63; 3.50)	-1.80	0.356
6-3	2.57	3.10	(6.50; 11.63)	0.83	0.930
7-3	14.13	3.10	(-11.53; 15.66)	4.56	0.002
8-3	-4.90	3.10	(-28.76; -1.57)	-1.58	0.479

Appendix E

Potassium content, mg/kg, 0-20 cm 1st production period

Difference of levels	Difference of means	SE of difference	95 % CI	T-value	P-value
1-3	-3.00	1.51	(-7.41; 1.41)	-1.98	0.251
2-3	7.90	1.51	(3.49; 12.31)	5.22	0.001
4-3	-4.00	1.51	(-8.41; 0.41)	-2.64	0.082
5-3	1.27	1.51	(-3.14; 5.67)	0.84	0.901
6-3	-2.37	1.51	(-6.77; 2.04)	-1.56	0.458
7-3	2.93	1.51	(-1.47; 7.34)	1.94	0.268

Potassium content, mg/kg, 20-40

cm Difference of levels	Difference of means	SE of difference	95 % CI	T-value	P-value
1-3	-13.60	1.58	(-18.19; -9.01)	-8.63	0.000
2-3	-10.83	1.58	(-15.43; -6.41)	-6.87	0.000
4-3	-12.87	1.58	(-17.46; -8.27)	-8.16	0.000
5-3	3.13	1.58	(-1.46; 7.73)	1.99	0.249
6-3	-11.77	1.58	(-16.36; -7.17)	-7.46	0.000
7-3	-11.20	1.58	(-15.79; -6.61)	-7.10	0.000

Potassium content, mg/kg, 0-20 cm 2nd production period

Difference of levels	Difference of means	SE of difference	95 % CI	T-value	P-value
1-3	-0.37	4.39	(-13.21; 12.48)	-0.008	1.000
2-3	-1.13	4.39	(-13.98; 11.71)	-0.26	1.000
4-3	14.27	4.39	(1.42; 27.11)	3.25	0.026
5-3	18.93	4.39	6.09; 31.78)	4.31	0.003
6-3	10.60	4.39	(-2.24; 23.44)	2.41	0.130
7-3	-3.07	4.39	(5.07; 23.20)	-0.70	0.968
8-3	1.17	4.39	(-28.76; -1.57)	0.27	1.000

Potassium content, mg/kg, 0-20 cm 2nd production period

Difference of levels	Difference of means	SE of difference	95 % CI	T-value	P-value
1-3	0.27	3.90	(-11.14; 11.67)	0.07	1.000
2-3	17.70	3.90	(6.30; 29.10)	4.54	0.002
4-3	33.10	3.90	(21.70; 44.50)	8.49	0.000
5-3	39.83	3.90	(28.43; 51.24)	10.21	0.000
6-3	25.40	3.90	(14.0; 36.80)	6.51	0.000
7-3	32.07	3.90	(20.66; 43.47)	8.22	0.000
8-3	2.30	3.90	(-9.10; 13.70)	0.59	0.987

Appendix F

The mobile P_2O_5 and exchangeable K_2O expressed as P and K

Fertilizer applied	P (%)		K (%)	
	Soil depth (cm)			
1st Production year	**0-20**	**20-40**	**0-20**	**20-40**
Lactofermented mix and biochar	24.14	20.20	30.02	31.40
Lactofermented mix	22.48	22.05	33.36	26.90
Mineral fertilizer	23.75	22.97	30.57	32.73
Stored cattle manure	23.05	22.33	28.40	26.06
Stored faeces	23.05	21.11	29.06	26.50
Stored urine	23.53	29.33	31.30	26.73
Control	21.81	22.11	28.38	25.73
2nd Production year				
Lactofermented mix and biochar	24.13	24.30	34.61	28.20
Lactofermented mix	22.50	24.70	28.95	32.78
Mineral fertilizer	23.75	25.41	30.08	43.83
Stored cattle manure	23.53	24.99	30.50	37.82
Stored faeces	23.05	25.91	24.49	41.03
Stored urine	23.53	28.42	26.73	40.57
Control	21.80	21.72	21.72	27.31
Vermicomposted lacto-fermented mix and biochar	25.60	24.30	34.61	28.18

For equivalence calculation the mobile P_2O_5 and exchangeable K_2O are expressed as P and K; the proportion of P in P_2O_5 is calculated from P_2O_5 (31x2+16x5=142) being equivalent to 100%, and a molecule of P as 21.83%; for K_2O (40+40=80+16=96), similarly, one molecule of K is 41.7%.

Appendix G

Soil bulk density, 1st production period

0-20 cm

Difference of levels	Difference of means	SE of difference	95 % CI	T-value	P-value
1-3	0.0800	0.0355	(-0.0233; 0.1833)	2.26	0.161
2-3	0.0500	0.0355	(-0.0533; 0.1533)	1.41	0.553
4-3	0.0800	0.0355	(-0.0233; 0.1833)	2.26	0.161
5-3	0.0833	0.0355	(-0.0199; 0.1866)	2.35	0.137
6-3	0.3267	0.0355	(0.2234; 0.4299)	9.21	0.000
7-3	0.1700	0.0355	(0.0667; 0.2733)	4.79	0.001

20-40 cm

Difference of levels	Difference of means	SE of difference	95 % CI	T-value	P-value
1-3	0.0867	0.0334	(-0.0106; 0.1839)	2.60	0.089
2-3	-0.0233	0.0334	(-0.1206; 0.0739)	-0.70	0.952
4-3	-0.1233	0.0334	(-0.2206; -0.0261)	-3.69	0.011
5-3	0.1267	0.0334	(-0.0294; 0.2239)	3.79	0.009
6-3	0.0000	0.0334	(-0.0972; 0.972)	0.00	1.000
7-3	-0.0067	0.0334	(-0.1039; 0.0906)	-0.20	1.000

Soil bulk density, 2nd production period

0-20 cm

Difference of levels	Difference of means	SE of difference	95 % CI	T-value	P-value
1-3	0.0433	0.0350	(-0.0589; 0.1456)	1.24	0.704
2-3	0.0433	0.0350	(-0.0589; 0.1456)	1.24	0.704
4-3	0.1000	0.0350	(-0.0022; 0.2022)	2.86	0.057
5-3	0.1967	0.0350	(0.0944; 0.2989)	5.63	0.000
6-3	0.0333	0.0350	(-0.0689; 0.1356)	0.95	0.875
7-3	0.1700	0.0350	(0.0678; 0.2722)	4.86	0.001
8-3	0.0067	0.0350	(0.0956; 0.1779)	1.37	0.619

20-40 cm

Difference of levels	Difference of means	SE of difference	95 % CI	T-value	P-value
1-3	0.0567	0.415	(-0.0646; 0.1779)	1.37	0.619
2-3	0.0433	0.415	(-0.1466; 0.0976)	1.24	0.990
4-3	0.1000	0.415	(-0188; 0.2612)	-0.56	0.020
5-3	0.1967	0.415	(0.0944; 0.2989)	5.63	0.002
6-3	0.0333	0.415	(-0.0689; 0.1356)	0.95	0.020
7-3	0.1700	0.415	(0.0678; 0.2722)	4.86	0.015
8-3	0.0067	0.415	(0.0956; 0.1779)	1.37	1.00

This work was performed at IHE, Delft Institute for Water Education (The Netherlands) and the Research Institute for Field Crops, Selectia, Balti (Moldova). The field work was carried out in village Boldureşti, Nisporeni district (Moldova). This research was funded by the Nuffic Fellowship Program (grant CF1080). The laboratory and field experimental work was also partly covered by the Laboratory of Hydrobiology and Ecotoxicology of the Institute of Zoology (Academy of Sciences of Moldova) and WiSDOM association.

Netherlands Research School for the
Socio-Economic and Natural Sciences of the Environment

D I P L O M A

For specialised PhD training

The Netherlands Research School for the
Socio-Economic and Natural Sciences of the Environment
(SENSE) declares that

Nadejda Andreev

born on 4 January 1972 in Nisporeni, Moldova

has successfully fulfilled all requirements of the
Educational Programme of SENSE.

Delft, 29 September 2017

the Chairman of the SENSE board

Prof. dr. Huub Rijnaarts

the SENSE Director of Education

Dr. Ad van Dommelen

The SENSE Research School has been accredited by the Royal Netherlands Academy of Arts and Sciences (KNAW)

K O N I N K L I J K E   N E D E R L A N D S E
A K A D E M I E   V A N   W E T E N S C H A P P E N

The SENSE Research School declares that **Ms Nadejda Andreev** has successfully fulfilled all requirements of the Educational PhD Programme of SENSE with a work load of 35.3 EC, including the following activities:

<u>SENSE PhD Courses</u>

o   Environmental research in context (2012)
o   Research in context activity: Initiate and organize the workshop "Perspectives of the reuse of human excreta from ecological sanitation for restoration of soil fertility in Moldova" (2014)

<u>External training at a foreign research institute</u>

o   Training on soil sampling methods and analysis, Selectia Research Institute of Field Crops, Balti, Republic of Moldova (2013-2014)
o   Online course Scientific and professional publishing on environment and sustainability, Open University Heerlen (2014-2015)
o   Environmental problems: crossing boundaries between science and society, Open University Heerlen (2014-2015)

<u>Management and Didactic Skills Training</u>

o   Supervising BSc student with thesis entitled 'The potential for the use of combined treatment of lacto-fermentation and vermi-composting of human waste for the application in agriculture' (2014)
o   Co-organising the international conference 'Environmental Challenges in Lower Danube Euroregion', Galati, Romania (2015)

<u>Oral Presentations</u>

o   *A concept for a sustainable sanitation chain based on the semi centralised production of terra preta for Moldova*, 4th International Dry Toilet Conference, 22-24 August 2012, Tampere, Finland
o   *Production of terra preta soil improvers by lacto-fermentation of organic residues*, International conference 'Rational use of natural resources - the basis for sustainable development', 10-11 October 2013, Balti, Republic of Moldova
o   *The effect of a terra preta-like soil improver on the germination and growth of radish and parsley*, 1st International Terra Preta Sanitation Conference, 28-31 August 2013, Hamburg, Germany
o   *Terra preta nova production for resource oriented excreta management in separately collecting sanitation facilities*, 3rd International Water Association (IWA) Development Congress & Exhibition, 14-17 October 2013, Nairobi, Kenya
o   *Processing of faeces from urine diverting dry toilets by combined biological processes: lactic acid fermentation and vermi/thermal composting*, International Conference on Terra Preta Sanitation and Decentralised Waste Water Systems, 18-21 November 2015, Goa, India

SENSE Coordinator PhD Education

Dr. ing. Monique Gulickx